The Climates of Canada

Contents

D0131487

T1-BKW-804

Author: **David Phillips**

Editors: **David Francis, Don Gullett, Ed Truhlar**

Managing Editors: **Nancy Cutler, John Sandilands**

Reviewers:
Edmonton: Neil Parker, Fredericton: John Dublin
Halifax: John Merrick
Montréal: Gilles Desautels, Jacques Miron, Suzanne Roy, Gérald Vigeant
Ottawa: Phil Coté, David Mudry, Regina: Ron Hopkinson, Ken Jones
St. John's: Stu Porter
Saskatoon: Les Welch
Toronto: Ross Brown, Ron Crowe, Ken Daley, David Etkin, Bruce Findlay,
  Don Gullett, Henry Hengeveld, Bill Hogg, Paul Louie, Joan Masterton,
  Barrie Maxwell, Bob Morris, Linda Mortsch, Michael Newark,
  Gary Pearson, Bryan Smith, Morley Thomas, Hans VanLeeuwen
Vancouver: Earl Coatta, Kirk Johnstone
Winnipeg: Barrie Atkinson, John Bendell, Rick Raddatz

Computer Support: Malcolm Geast, Michael Webb

Clerical Support: Marvena Voss

Typing: Pearl Burke, Una Ellis

Production Liaison: Albert Wright

Translation: Jean Giguère, Danielle Lagueux, Manon Lavoie, Roger Martel,
  Ann McLean, Jacques Mercier, Claude Paré, Daniel Pokorn,
  Caspar von Schoenberg

French Editing: Claude Paré, Gilles Tardif

Design: Galer + MacMillan Communications Inc.

Computer Graphics: Hao Le

Cartographie: Galer + MacMillan Communications Inc.

Photo Credit: Supply and Services Canada - Photo Centre,
  unless otherwise indicated

Illustrator: Wayne Terry

Printer: Johanns Graphics

Dedication:
This book is dedicated to Morley Thomas who has contributed enormously to the history, science, and service of Canadian meteorology for nearly half a century.

The Climates of Canada

> **"Having learned to endure in a climate of vast extremes, we Canadians make the most of its small mercies; a winter that arrives a week or two late, a summer that comes a few weeks early, a cloudless day in February, a snowcover that is a few centimetres less than the year previous."**
>
> *Val Clery*
> *"Seasons of Canada"*

Foreword

Canadians have a profound and abiding obsession with their climate. Indeed, Canadians are among the most climate-conscious, weather-conversant people in the world. Why? Because Canada is a land of climatic contrasts and extremes, and in one way or another its climatic diversity touches and influences many aspects of our lives.

The diversity of the Canadian climate is evident in the number and variety of Canada's climatic landscapes — permanently frozen ice-caps, windswept treeless tundra, luxuriant Pacific rain forests, hot semi-arid scrub lands, polar deserts, and sun-drenched grain fields. It is evident too in the changing seasons and in the daily alternations of the weather. The contents of a typical Canadian wardrobe — parkas, raincoats, rubbers, ear muffs, Bermuda shorts, umbrellas, scarves, boots or mukluks, leg warmers, tuques, and bikinis — are a reminder of the many dimensions of our climate.

There is something unpredictable about the Canadian climate as well. No season or year can be counted on to give an accurate indication of what the next one will be like. Cool summers may follow warm winters or vice versa. Whereas some regions benefit from timely rains, extra warmth, and a late freeze, a not-too-distant region may endure severe drought, disastrous hailstorms, or a surprise snowfall, and their situations may be reversed with bewildering suddenness.

Weather is always the first topic of conversation among friends or strangers alike. This is not only because it is an easy or traditional opener, but because it is truly important. At one time or another, weather can be a benevolent resource, a terrifying hazard, a challenge, a threat, a pleasure, or a disappointment. In the course of an average year, Canadians will probably experience it in all these guises. Most of us, in fact, will look back with pleasure and even pride at some of the more bizarre weather we have endured, although our responses at the time may not have been so enthusiastic.

The climate has its economic significance too. In a typical winter, Montréal may spend $40 million on snow removal whereas Victoria may spend only a few thousand. For farmers and ski-lodge operators, the whims of the weather can spell the difference between prosperity and bankruptcy. Because of the extremes of our climate we are also among the world's largest energy users and probably the world's most enthusiastic developers of indoor and underground shopping centres.

Even though Canada's climates are diverse and demanding, they are by no means the most extreme that nature can offer. Canada holds none of the world's major weather records. Worse storms and greater extremes of cold, heat, wetness, and dryness occur elsewhere. Over the years, much has been written about our weather, the factors that shape it, and the impact it has on our lives.

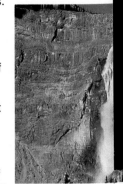

From Tuller's *Climate of Vancouver Island* to Peach's *Climate of Newfoundland and Labrador*, descriptions of the climate of Canada and its regions abound. However, two of the best surveys, Boughner and Thomas's *Climate of Canada* and Hare and Thomas's *Climate Canada*, are sadly out of print. This book has been launched largely to fill the gap left by their absence. The author acknowledges his extensive reliance on their material as well as on the other published sources listed in the references near the end of this publication.

The book is divided into four sections. The introductory section is intended to acquaint readers with those factors that shape and control Canada's climates. The second section looks at the elements — temperature, rain and snow, ice, storms, floods, and other components of our weather. The third section surveys the climates of each of the provinces and territories. The fourth is a short section on climate change. These are followed by a glossary, a list of references, and a selection of tables detailing averages and extremes for 17 weather stations in Canada — sites representative of the principal climatic regimes. This material does not necessarily have to be read in sequence. Readers are free to hop about from topic to topic as their interests or needs dictate.

This is not a reference for scholars and decision makers. Instead, it is meant for anyone seeking a general description of the diversity of Canada's climates and explanations why such diversity exists. For those who wish to study these climates in more detail, there is plenty of published information available through the Atmospheric Environment Service of Environment Canada.

It is my hope that readers will find this an entertaining as well as a useful introduction to the climates of Canada and that some of them at least will be stimulated to seek out and read the many more-detailed studies that are available.

David Phillips

"As a people we are chained to the mystery of our endless sky, to the sudden flooding rush of spring, the fat buzz of summer, and the ruthless death of winter through which in every crack of ice we see the green promise of a mystical tomorrow."

Harold Town
Prologue, Land of Land in
"Canada With Love"

Climate Controls

Every place on Earth, whether a Newfoundland fishing bank, an Amazon rain forest, or a Montréal roof-top, experiences a distinctive and fairly regular pattern of weather from year to year. Of course, the pattern is not precise, and there is variation from one year to another, but generally the amount of sunshine and rain, the range of temperatures, and other features of the weather tend to be similar from one year to the next. It is this "average weather" that we call climate.

Weather is caused primarily by the exchange of energy between the sun, the earth's surface, and the atmosphere. The sun supplies the energy, free of charge and in endless amounts. The earth's shape and motion determine how much sunlight and heat are received and where. The atmosphere acts as a great insulator and equalizer — insulating the earth from extraterrestrial extremes and balancing the pools of heat and cold between the equator and the poles. Surface water, the shape of the land, and the nature of the ground cover also have an important influence on weather patterns.

The complex meshing of these various factors determines the character and diversity of a place's climate.

LLOYD FREESE

THE SUN AND THE EARTH

Energy From The Sun

The sun's energy travels through nearly 150 million kilometres of space before arriving at the outer edge of the earth's atmosphere. Only about half of that energy actually reaches the earth's surface, however. The rest is reflected back to space by clouds, scattered by dust and water vapour, or absorbed by the atmosphere and re-radiated.

At the surface, still more of the incoming radiation is reflected back to the atmosphere and space, but part of it is absorbed by the land and oceans. The colour of a surface affects how much energy it will reflect or absorb—the lighter its colour, the more energy it will reflect; the darker its colour, the more it will absorb. A field covered by snow will reflect 60-80% of the radiation striking it. The same field covered by crops absorbs all but 20% of the incoming radiation.

When it is absorbed, the solar energy is converted to heat to warm the earth and the overlying air, to evaporate water, or to melt snow. Some of this heat escapes to outer space, but most is absorbed by water vapour, carbon dioxide, and other gases in the atmosphere and radiated back to earth. The atmosphere, in fact, works like a greenhouse window, letting in copious amounts of energy from the sun but allowing only a portion of it to escape immediately back to space. This phenomenon, known as the greenhouse effect, is essential to life on earth. Without it, our planet would be as cold and barren as the moon.

Eventually, however, all of the energy which the earth receives from the sun returns to outer space. As a result, the earth neither gains nor loses energy, and thus, over the long run, it neither heats up nor cools down.

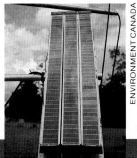

ENVIRONMENT CANADA

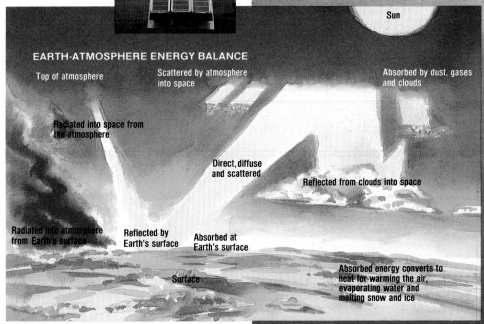

EARTH-ATMOSPHERE ENERGY BALANCE

Sun

Top of atmosphere

Scattered by atmosphere into space

Absorbed by dust, gases and clouds

Radiated into space from the atmosphere

Direct, diffuse and scattered

Reflected from clouds into space

Radiated into atmosphere from Earth's surface

Reflected by Earth's surface

Absorbed at Earth's surface

Surface

Absorbed energy converts to heat for warming the air, evaporating water and melting snow and ice

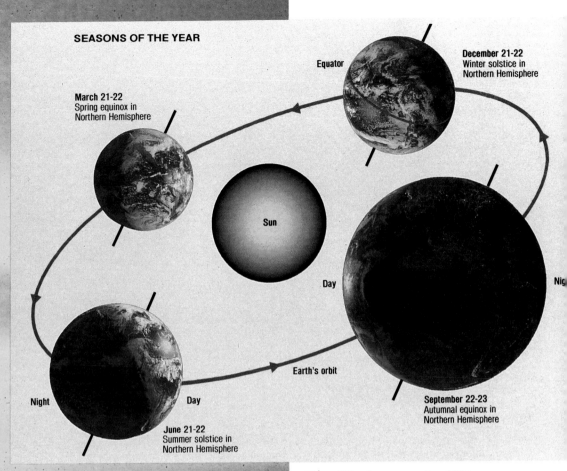

SEASONS OF THE YEAR

March 21-22
Spring equinox in
Northern Hemisphere

Equator

December 21-22
Winter solstice in
Northern Hemisphere

Sun

Day

Nig

Earth's orbit

Night   Day

June 21-22
Summer solstice in
Northern Hemisphere

September 22-23
Autumnal equinox in
Northern Hemisphere

Two Motions and a Tilt

All parts of the earth do not receive the same amount of energy from the sun. As a result, some parts heat up more than others. This unequal heating sets the atmosphere in motion: the heated air expands and rises over the warmer regions and heavier, cooler air moves in from the colder regions to replace it. The atmosphere, in effect, tries to distribute the heat more evenly.

To understand why the sun's energy is distributed unevenly, we must look at the way the earth moves in relation to the sun. This movement can be described as two motions modified by a tilt. The first motion is the rotation of the earth on its axis, spinning in an easterly direction from day to night and back to day every 24 hours. The second motion is the movement of the earth around the sun. Moving along an egg-shaped orbit, the earth makes one complete revolution around the sun in a year.

The tilt is a 23 1/2° tip of the earth's axis. While the earth travels around the sun, this tilt determines the lengths of night and day for each latitude. It also determines the angle at which the sun's rays strike the

OBLIQUE AND DIRECT SUN'S RAYS

The sun's rays at the equator are mostly perpendicular and therefore intense. Over Canada the solar beam passes through more atmosphere where it is absorbed or re-flected, and strikes the surface as an oblique ray. Because it is spread over a greater surface area, it is less concentrated.

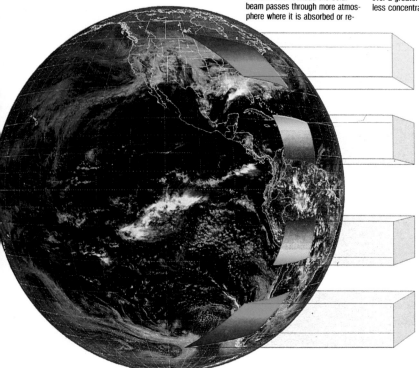

an ever increasing angle. The beam therefore becomes more spread out and takes a longer path through the atmosphere, thus losing more energy on its way to the earth's surface.

In summer, however, the decreased intensity of the sun's radiation at the higher latitudes is balanced by the greater length of the days there. Thanks to its longer days, northern Canada receives about the same amount of solar energy in June as Florida.

Because of its influence on the amount of solar radiation received, the north-south position of a place has a significant effect on the intensity of climatic events. It also determines the type and frequency of weather systems to which a region is exposed. Generally, higher latitudes mean lower temperature, less sunshine, radiation, and evaporation, and greater relative humidity, cloudiness, and snowcover.

Only about half of the solar energy that reaches the earth's atmosphere actually penetrates to the surface. Much of it is reflected by clouds and returns to space. Some of it is scattered by dust particles and water vapour to produce diffuse radiation, a portion of which reaches the earth's surface. Some is absorbed by the atmosphere and re-radiated.

Twice each year, at the spring and fall equinoxes, the noon sun is directly overhead at the equator. At the spring equinox (March 21), day and night are of equal length. Then the days gradually grow longer until June 21 (the summer solstice), when the sun is directly overhead at noon above the tropic of Cancer and the area north of the Arctic Circle has 24 hours of continuous daylight. After the summer solstice, the days again grow shorter. By September 21 (the fall equinox) day and night are again equal. The days continue to shorten until December 21 (the winter solstice), when the noon sun is directly above the tropic of Capricorn and darkness occurs around the clock north of the Arctic Circle.

surface of the earth at a particular time and thus the amount of energy that the surface will receive.

Without the tilt, the sun would shine directly over the equator all year round and there would be no seasons. But because of the tilt, the sun shines directly over the equator only twice a year. The rest of the time the sun is directly overhead somewhere between the tropic of Cancer in the Northern Hemisphere and the tropic of Capricorn in the Southern Hemisphere. Each of the tropics is 23 1/2° away from the equator—the same number of degrees as the tilt of the earth's axis. Consequently, one hemisphere gets more direct solar radiation for part of the year than the other. The result is the seasons. When the sun is north of the equator, between March 21 and September 21, the Northern Hemisphere experiences its warmest weather and the Southern its coldest. When the sun is south of the equator, between September 21 and March 21, the opposite happens.

Because the earth follows an egg-shaped orbit, its distance from the sun changes during the year. But, surprisingly, this change in distance has nothing to do with causing the seasons. In January, the sun is actually 8 million kilometres closer to the earth than in July.

The North-South Effect

The latitude of a place also determines what share of the sun's radiation it will receive. The largest amount of solar radiation is received at or near the equator, since it is here that the sun's rays are closest to being perpendicular to the earth's surface. As a result, the rays are more concentrated when they reach the surface, and less of their energy has been lost in passing through the atmosphere. As we move towards the poles, however, the solar beam passes through the atmosphere at

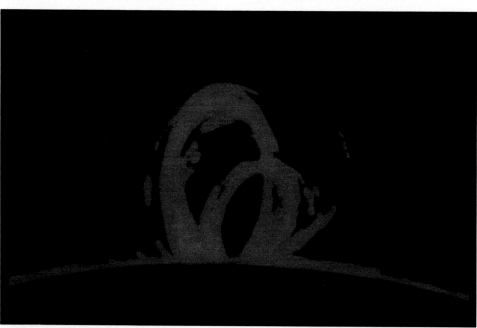

THE MOVING ATMOSPHERE

Each year Canada loses more heat to outer space than it receives from the sun. Despite this loss, our climate does not get continually colder because over the globe the atmosphere and ocean currents redistribute heat from the tropics to the poles. While air in the tropics is heated by surface contact, it rises and moves towards the poles. At the same time, heavy, cold polar air moves towards the equator to replace it. The earth's spin deflects these movements, setting up a system of global wind belts and pressure cells. As a result of this deflection, atmospheric currents in the Northern Hemisphere generally move from west to east rather than directly north and south.

Most of Canada lies within the belt of latitude known as the zone of the westerlies.

Here prevailing winds are from the west (winds are named after the direction they blow from), and weather patterns tend to move from west to east. Other Northern Hemispheric winds include the northeast trades in the tropics and the polar easterlies in the Arctic. Unlike the westerlies, however, the easterlies do not have a significant effect on Canada's climate. The trade winds lie too far to the south, and the polar easterlies affect only the Arctic, where their impact is moderated by a variety of local factors.

Moving around within the westerly stream are large migratory pressure systems associated with outbreaks of cool or cold Arctic air from the north and surges of mild or hot, humid air from the south. Low pressure areas (called cyclones) and their associated fronts are usually responsible for cloudy, windy, and wet weather. High pressure areas (anticyclones) generally produce clear, dry weather.

Air Masses

Newspaper headlines such as "Frigid Air Grips Prairies" or "No Relief from Ontario Heat Wave" show that weather conditions may be similar over a wide area. The conditions are similar because the entire area is covered by the same mass of air. An air mass is defined as a large section of the lower atmosphere with a horizontal uniformity of temperature and moisture conditions. If cold, dry, Arctic air moves into the Maritimes, the weather will stay cold and dry until the northern air mass is replaced by a warmer, moister type of air from the west or south. The transition between two contrasting air masses may be relatively rapid. In summer sudden changes of air mass may be signalled by thunderstorms or rain showers.

Air masses meeting over Canada may have their origins thousands of kilometres away. They may develop over the frozen Arctic, the cool waters of the North Pacific Ocean, the tropical seas of the Caribbean, the desert lands of the American Southwest

or the iceberg-infested waters of the North Atlantic. Over these source regions, the air acquires uniform temperature and moisture conditions, taking these characteristics from the area over which it forms. An air mass may stagnate for days or weeks at a time before the upper-air circulation causes it to move.

Along the way, the air mass is changing continually. By mixing with other air masses in a cyclonic (counter-clockwise) motion and by passing over varied surfaces — from snowcover to bare ground or from ocean to land — air streams swirling over Canada rarely retain the properties of their original source region for more than a few days.

Cold Air: In winter most of northern Canada and the Arctic Ocean make up one huge snow and ice-covered surface. This area is the source of bitterly cold air that frequently penetrates most of Canada except coastal British Columbia and the outer coasts of Nova Scotia and Newfoundland, but even on occasion those localities as well. Before moving south, the air may sit still for weeks under the prolonged darkness of winter, giving uniformly ice-cold, dry, clear, and stable weather to large portions of the Arctic.

Summer brings rapid changes. Additional solar heat and the resulting melting of snow and ice leaves the white surface covered with pools of water. The ice-pack begins to break up, exposing more open water, and the source region of cold Arctic air shrinks, with the result that the Arctic air mass takes on more of the character of ocean air.

Tropical Air: The warm Caribbean Sea and adjacent waters are source regions for humid tropical air. Although the southern air mass is a rarity in winter in Canada, it occurs frequently enough in summer in its original form to give parts of central and eastern Canada tropic-like weather — hot, hazy, and humid.

Dry tropical air from the southwestern American deserts sometimes enters western and central Canada, but only rarely.

Wet Air: Maritime or ocean air is common in Canada at all times of the year, although its properties vary widely from place to place. It usually begins as Arctic air, but it derives its maritime character after spending considerable time over the northern oceans.

In winter, North Pacific air arrives along the Pacific coast from the west, accompanied by rain and thick layers of cloud. Air streams that manage to make it over the coastal mountains arrive in the continental interior much drier than the air along the coast and much milder than the Arctic air

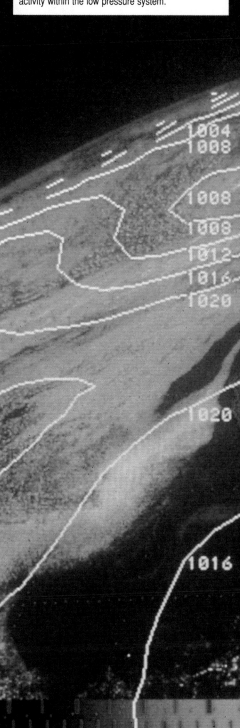

1004
1008
1008
1008
1012
1016
1020
1020
1016

normally sitting over the Prairies. In summer, Pacific air enters Canada accompanied by cool to mild, dry, and bright weather.

North Atlantic air does not usually penetrate far inland. In winter, Atlantic air streams bring mild, cloudy, drizzly weather to parts of Atlantic Canada and the Gulf of St. Lawrence. In summer, Atlantic air remains cool and moist, especially in the vicinity of the cold Labrador ocean current.

Travelling Weather Systems

Our weather is also affected by large cells of high or low pressure that are present almost permanently over certain parts of the globe.

Two large areas of high pressure, for example, are almost always found in the subtropical latitudes of the Pacific and Atlantic oceans. The Pacific High affects Canada more in summer than in winter. In the summer, it brings a northwest flow to the coast, blocking most storms from reaching the mainland. In fall and winter the high moves southward and weakens, no longer able to block the movement of storms to the coast.

In the Atlantic Ocean, the Bermuda or Azores High is a similar but smaller-scale feature. It moves considerably and in doing so is a strong determinant of the weather affecting eastern Canada. Some of the most intense hot spells in summer come under the control of the Bermuda High, as does the occurrence of Indian summer in the fall.

Complementing the two oceanic highs are two subpolar oceanic lows — the Aleutian Low in the North Pacific close to Alaska and the Icelandic Low in the North Atlantic northeast of Newfoundland. These lows are not permanent entities but usually slow-moving disturbances passing through these areas. Pressure patterns averaged over several years always indicate their presence. The Aleutian Low is a source region for many of the lows that enter Canada through the West Coast. The Icelandic Low is the final destination for many of the lows that have moved across North America.

A third high pressure cell is the Arctic High. It is highly variable in position and intensity but usually predominates on the Arctic mainland in spring.

The boundary (or front) between warm and cold air masses is a fertile breeding ground for low pressure systems. They are big features, extending 2000 to 3000 km across and are often fast-moving, hurrying along in the westerly circulation, crossing the continent in about three or four days. Large areas of poor weather — widespread cloud, precipitation, wind, and obscured visibility — usually accompany these low pressure systems.

CYCLONES AND ANTICYCLONES

To most of us, cyclones are ferocious revolving storms like hurricanes or tornadoes, but to meteorologists they are merely depressions or areas of low pressure associated with unsettled weather. The scientific and the common meanings of the term are not entirely inconsistent, however, since both hurricanes and tornadoes are particularly intense depressions with very low central pressures. The opposite of cyclones are anticyclones, and these are areas of high pressure, usually associated with clear, stable weather.

Pressure cells are a product of the motion of large air currents within the atmosphere. This motion creates centres of higher or lower pressure around which the surrounding air moves in a spiral pattern. In a low pressure area, the air spirals in an anticlockwise or cyclonic direction in the Northern Hemisphere. There is also rising and falling motion of air within a pressure cell — rising within a low pressure area and descending within a high.

The lowest air pressure ever recorded in Canada was 94.02 kilopascals (kPa) measured at St. Anthony, Newfoundland, on January 20, 1977. The highest was 107.96 kPa, measured at Dawson, Yukon, on February 2, 1989. Air pressure at the centre of Hurricane Gilbert in the Caribbean dropped to 85 kPa on September 14, 1988, the lowest central pressure ever recorded in the eye of a hurricane.

WINTER AIR MASSES AND CIRCULATION

- – – – Polar jet stream
- ▶ ▶ ▶ Primary storm tracks

● Continental Arctic
- very cold −25 to −50°C
- dry, very stable
- pronounced temperature inversion

● Maritime Arctic
- very unstable
- clouds, frequent showers or flurries
- visibility good except in showers

● Maritime Polar
- milder and more stable than Arctic air

● Pacific Maritime Tropical
- light winds, cooler than Atlantic air
- comes to North America from west or northwest
- stable in lower 1000 m (marine stratum)

● Atlantic Maritime Tropical
- comes to North America from south or southeast
- warm and humid

SST Sea surface temperature

SUMMER AIR MASSES AND CIRCULATION

- – – – Polar jet stream
- ▶ ▶ ▶ Primary storm tracks

● Pacific Maritime Tropical
- H pressure precludes moist air

● Atlantic Maritime Tropical
- oppressively hot and humid
- unstable, frequent thunderstorms

● Maritime Arctic
- continental air modified by open seas, lakes and swamps

● Maritime Polar
- warmer and more stable than Maritime Arctic air

● Continental Tropical
- hot, dry, unstable

SST Sea surface temperature

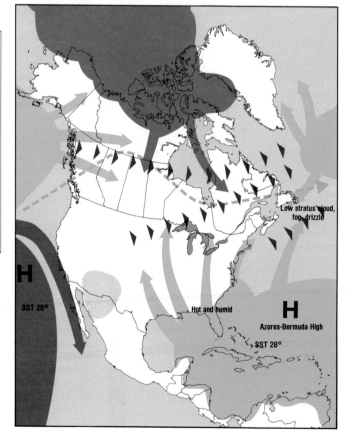

In winter, frontal weather is generally more disturbed and vigorous and more frequent, with storms every two or three days over southeastern Canada. A favoured path for low pressure systems takes them from the west and southwest along the St. Lawrence Valley and across Newfoundland. The Atlantic coast storm, a low pressure system which frequently forms over Cape Hatteras, is often intense and brings strong winds or gales and heavy precipitation in a multitude of forms. Merging Great Lakes-St. Lawrence storms and Atlantic storms in the Cape Hatteras region redevelop and intensify along the Atlantic seaboard, aided by the surface heat from the Gulf Stream.

During the winter, coastal British Columbia receives a constant procession of Pacific disturbances. Another frequent source area for lows is east of the Rocky Mountains, most often in Alberta. Here, new storms develop, and old ones regenerate and move eastward across the continent. The Alberta Low is normally not a large or vigorous disturbance, however, compared with the lows coming out of the American south.

In summer the large oceanic semipermanent highs dominate North America. The path of migrant lows is farther north, and storms are less intense than in winter and are about half as frequent. Over the Pacific, frontal storms strike the B.C. coast farther north and less often, while in central and eastern Canada storm activity is much less frequent. Wet summer weather comes from local showers or thunderstorms, while the land heats under the summer sun, as well as from storms embedded in low pressure systems.

Tropical hurricanes are also low pressure systems, but they are smaller and considerably more violent than their mid-latitude counterparts. They normally form in the Caribbean in the late summer or early fall, deriving their energy and moisture from the warm waters of the tropical ocean. In their trek northward over the relatively cooler waters of the North Atlantic, hurricanes tend to become larger, less intense, and less rainy. True hurricanes sometimes reach the Maritimes, but usually they merge with mid-latitude lows and rejuvenate before being pushed to the northeast Atlantic by the strong westerlies blowing off the continent.

High pressure cells also move in the westerlies. Associated with them is clear, dry weather. Alternating with lows, they add to the weather variability of the temperate latitudes. Occasionally, a high will stall and block storms from advancing or will divert the flow of air masses around it. Blocking generally produces weather extremes — intense heat or cold or long-lasting spells of dry or wet weather.

the shape of the land

If the general circulation of the atmosphere controls weather, it is mainly physical geography that fashions climate. The diversity of Canada's landscapes and the variety of its climates are not unrelated. From towering mountains, plains, lowlands and highlands, one might expect a full range of climates, from arid to moist and from maritime to continental. Given a different landscape — say a west-east mountain chain in the Prairie provinces, a youthful Appalachia, or a mature Cordillera — Canada's climates would be much different than they are today.

Topographical features — elevation, slope, exposure, and orientation of landforms — combine to alter greatly the regional and local patterns of climate. Elevation, for example, can have a number of climate consequences. Atmospheric pressure decreases with altitude and therefore, with certain exceptions, does temperature. One of these exceptions is an inversion, a phenomenon in which cold air is trapped near the ground by a layer of warmer air overhead. Another common exception occurs in valleys, depressions, or other low-lying areas, where cool night air drains downward and settles into the bottom creating a "frost hollow." White River, Ontario, which sits in such a hollow, has an average of 309 frost days a year, almost double that of nearby stations on elevated or level terrain. Another consequence of the local topography at White River is the occurrence of calm or very light winds almost 50% of the time in winter.

Other weather elements generally increase with elevation. The amount and frequency of precipitation is greater at higher elevations on the windward side of a mountain range. Sometimes humidity increases when clouds are encountered. Lightning and cloud cover are also more frequent. Wind velocity strengthens with height too, although other local factors such as the slope of the ground or the presence of obstacles may change its speed or direction.

LLOYD FREESE

Canada's Landscapes

Air streams moving through Canada encounter a variety of major landforms. Some of these have very large-scale effects on our climate. Others have only local effects.

The Canadian landscape has six major relief forms. The most dramatic of these is the Western Cordillera, part of the complex system of mountains, plateaux, and valleys which stretches from Alaska to Mexico and beyond. It is a massive barrier that stops the low-level, moist, cloudy Pacific air streams from entering Canada easily. While the air is forced aloft over successive mountain ranges, it gives up its moisture and becomes relatively dry and warm by the time it sweeps into Alberta. Were it not for the Cordillera, the maritime climate of the Pacific coast would extend well inland. The Cordilleran barrier also protects the west coast from occasional outbreaks of cold Arctic air.

Moving east we come to the Great Plains or the Prairies, extending from the Arctic coast to the American border and from the foot of the Rockies in Alberta to the Shield country of eastern Manitoba and northern Ontario. This broad, flat corridor, running northwest to southeast, is devoid of high mountains and is open to invasions of cold Arctic winds in winter and hot, dry, southern air in summer. Consequently, the

Prairie Provinces, especially in the agricultural south, witness some sudden weather changes and incredible temperature ranges of over 50°C in an average year.

Central Canada is dominated by the ancient rock surface of the Canadian Shield, with Hudson Bay in its sunken centre. Shield country is strewn with lakes, bogs, and rocky outcrops. Despite its ruggedness, the Shield has little effect on the movement of large air streams across it. Its topography has some effect, however, on local precipitation and temperature.

The Appalachian Highlands are an eroded plateau, about 300 to 400 metres above sea-level. The relief is broken up by valleys and by higher ridges running from southwest to northeast. Extending north from the United States near Lake Champlain, they extend eastward along the south shore of the St. Lawrence through Quebec, continue into New Brunswick and Nova Scotia, and turn up again in Newfoundland. Because the highlands are aligned with the normal path of passing weather systems rather than at right angles to it, they have less effect on precipitation than they otherwise would. Also, since the atmosphere flows mainly from west to east across North America, the location of the Appala-

chians at the eastern end of the continent minimizes their overall effect on the climate.

The Great Lakes-St. Lawrence Lowlands extend from southern Ontario to near Québec City, where the Shield and Appalachian systems converge. The relief of the Lowlands is subdued, and apart from local precipitation and the usual effect of altitude on temperature, they do not have a significant effect on regional climate. They are, however, a favoured path for storms crossing North America from the southwest.

The Arctic Archipelago, the large triangle of islands lying north of the Shield, is not a major control on Canada's climate. Locally, however, the topography of the larger eastern islands, such as Ellesmere, Devon, and Baffin, has an effect on precipitation. Mountains on these glacierized islands rise some 2000 to 2500 metres above sea-level. Since the prevailing atmospheric flow in the high Arctic is from the east, the eastern flanks of these islands are therefore wetter than the western, which lie in the precipitation-shadow of the mountains.

OROGRAPHIC EFFECT

A — Cooling rate 0.5 to 0.8°C for every 100-m rise in elevation

B — Warming rate 1.0°C for every 100-m drop in elevation

Prevailing winds

Cold

Cool — wet

Warm — dry air

B

Chinook winds

A

Dry — warm

Moist maritime air lifted

Cool — wet

Rain shadow — dry

Mild

Land

Ocean

RAIN-SHADOWS

From Hope in the lower Fraser Canyon to Ashcroft on the North Thompson River is only a three-hour drive along British Columbia's Highway 1. But in the course of that trip the vegetation changes dramatically from the lush verdure of a coastal rain forest to the dry grass and sagebrush of a semi-desert. It is the kind of contrast one would expect to find only between places thousands of kilometres apart. Its occurrence in such a short distance is explained by what climatologists call the orographic effect—the tendency of air to lose moisture with increasing elevation. The effect is commonly seen in mountainous areas, where rain falls frequently on the windward side of the mountains and the leeward side forms a rain-shadow in which precipitation is much sparser.

We can see how the orographic effect works by following the prevailing westerlies as they move off the Pacific and cross the mountains of British Columbia. These winds, heavily laden with moisture, begin to rise when they encounter the Coast Mountains. While they move upwards, the air cools at a rate of 0.5 to 0.8°C for every 100-metre increase in elevation. The exact rate of cooling depends on the amount of moisture present. As a result of this cooling process, clouds, fog, and abundant precipitation characterize the weather along the windward face of the mountains.

Once the mountain crest is crossed and the air descends the leeward slopes, increasing pressure causes it to heat up. It gains heat at a rate of 1.0°C or more per 100 metres, slightly higher than its rate of cooling during ascent. In this rain-shadow area, water droplets evaporate, clouds break up, and surface drying occurs.

Not all the moisture is wrung out of the air on its windward ascent. Rain and snow spill over to the leeward slopes and valleys, but in lesser amounts than on the opposite side of the mountain at the same altitude and in decreasing amounts as the air moves away from the coast.

OCEANS AND

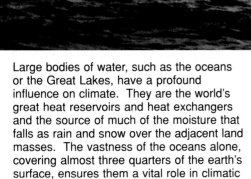

Large bodies of water, such as the oceans or the Great Lakes, have a profound influence on climate. They are the world's great heat reservoirs and heat exchangers and the source of much of the moisture that falls as rain and snow over the adjacent land masses. The vastness of the oceans alone, covering almost three quarters of the earth's surface, ensures them a vital role in climatic processes.

Water heats up more slowly than land and holds its heat longer. As a result, it has a moderating effect on climate. Generally, continental air is warmer in summer and colder in winter than ocean air at the same latitude. In addition, the times of seasonal temperature extremes are delayed by three to six weeks or more over deep, ice-free bodies of water.

Large water bodies are generally precipitation sources, wind generators, and cloud enhancers when they are warmer than the air above them. However, when the water is colder than the air, precipitation is suppressed, winds are reduced, and fog banks are formed.

Ocean Currents

Ocean currents are major movers of heat

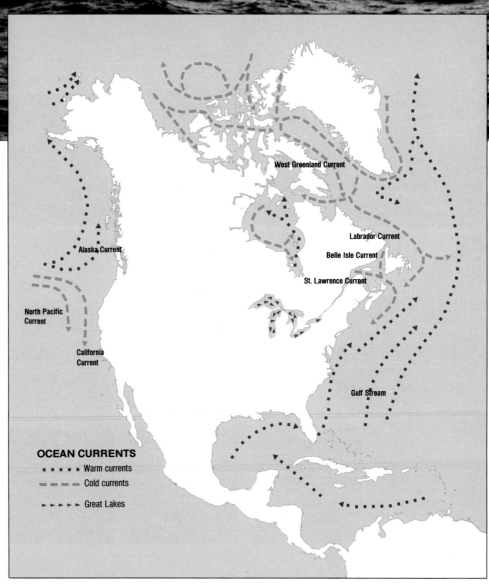

West Greenland Current

Labrador Current

Belle Isle Current

St. Lawrence Current

Alaska Current

North Pacific Current

California Current

Gulf Stream

OCEAN CURRENTS

· · · · · Warm currents

– – – Cold currents

► ► ► ► Great Lakes

LAKES

and cold around the world. The Gulf Stream is the warm current that flows northward along the Atlantic coast of North America before turning west and flowing towards Europe. It has a much greater effect on the European climate than on eastern Canada's, however, because of the prevailing westerly air flow. The other major ocean current on our Atlantic coast is the Labrador Current. Frigid and ice-filled, it flows south from the Arctic and collides with the Gulf Stream off the southeast coast of Newfoundland, producing extensive banks of fog.

The Alaska Current in the Pacific is a warm current but is much weaker than the Gulf Stream. Its moderating influence is confined to a narrow strip of the Pacific coast because of the barrier effect of the Coast Mountains.

Pacific Ocean

Since much of our weather comes from the west, the Pacific Ocean plays a significant role in shaping our climate. However, the true oceanic climate does not extend far inland, because the north-south alignment of the Cordillera impedes the eastward penetration of Pacific air. The generally warm and always ice-free waters of the Pacific bring mild winters and cool summers to the British Columbia coast, giving its climate a quality of unusual moderation for such high latitudes. Pacific air dominates the country for much of the year, but its moisture properties are greatly changed once it crosses the Cordillera.

Great Lakes

With a volume of 23,000 cubic kilometres, the Great Lakes act as a vast heat reservoir for modifying the weather of southern Ontario and adjacent Quebec. In winter, cold air is warmed several degrees by the open water, and the atmosphere is fed large amounts of moisture, some of which is returned as lake-effect snow downwind of the lakes. In spring, mid-lake waters are colder than they were in winter, and passing moisture condenses above the lake giving rise to shallow fog. In summer, comforting onshore lake breezes blow on most days, and in fall the stored heat of summer is released slowly enough to delay the first frost by one or two weeks within 10 to 15 km of the shoreline.

Other large inland lakes, such as Lake Winnipeg, Great Bear Lake, and Great Slave Lake, modify the climate of adjacent shores but not to the extent of the Great Lakes, which remain open longer and have more of a warming effect than the large northern lakes.

Hudson Bay

Hudson Bay covers almost as much area as all the lakes in Canada. This immense inland sea penetrates deep into the centre of Canada, but coastal areas receive little warmth from its cold waters. The bay is frozen almost completely from late December until late June, and during the brief summer, melting ice keeps surface temperatures near freezing. Fog is most frequent over the bay in summer, when warm air cools while passing over the colder waters. In fall the bay emits heat and moisture, adding rain and snow, cloud, and strong winds to the rawness of the coastal climate.

Atlantic Ocean

The climate of the Atlantic Provinces is less maritime than that of the Pacific coast because prevailing winds are off the continent. When they do blow onshore, their path brings them across the cold, iceberg-infested waters of the Labrador Current. Onshore winds, common only in spring and summer, reach the coast laden with moisture, producing fog and low cloud. In winter, an occasional incursion of easterly air breaks up the pack ice and provides a spell of milder temperatures and higher humidities. Unlike Europe, eastern Canada does not benefit from warm Atlantic Ocean currents.

Arctic Ocean

The Arctic Ocean contains enormous fields of permanent pack ice and drifting ice floes. At the height of summer, only about 10% of the ocean is free of ice. In winter, the Arctic surface is largely ice-covered. Consequently, its effect on passing air streams is more like that of a mass of land, and it has little moderating influence on the weather. Heat conducted through the ice, however, keeps winter temperatures on the ice from reaching the extremes that often occur over land. Ice fog and arctic sea smoke form over breaks in the ice-pack. During the brief ice-melt season — June, July, and August — the very cold waters refrigerate the North and produce extensive sea fog and layers of cloud, especially in the eastern Arctic. In fall, just before freeze-up, cold Arctic air comes into contact with the now relatively warm seas and sets up cumulus clouds, frequent snow flurries, and poor visibility.

THE CITY
& the Country

Soils, rocks, vegetation, buildings — all impart a different texture to the surface of the landscape and all influence climate to some degree. The effect is usually local rather than large-scale, but it can still be dramatic.

The fact that different kinds of soils vary in their ability to absorb and retain heat can have a measurable effect on frost occurrence and the length of the growing season. Sandy areas, for example, tend to be a degree or so warmer by day than areas where the soil is predominantly clay, although the air temperature at night tends to fall lower by 3 or 4°C over sand. Light-coloured surfaces reflect more heat than dark surfaces, and wet soils have less of a temperature swing from day to night than dry soils.

Vegetation can also influence local climates, with forests producing greater effects than other, more open, types of plant cover. Forests can modify climate in several ways, but generally humidities and night-time temperatures are higher under a forest canopy, with wind speeds and daytime temperatures being lower. The amount of sunshine, snow, and rain is also less within a forest, and climatic extremes are reduced.

Much of the climate modification under the forest canopy depends on the species, height and density, and age of the trees, and on the climate regime, season of the year, and local geography.

HEAT ISLANDS AND URBAN DOMES

Most large cities serve as gigantic radiators warming up the outside air. Office buildings, factories, and cars spew out enormous quantities of waste heat. As well, brick, concrete, and asphalt surfaces absorb and store heat during the day. At night the stored heat drifts from underneath or behind these surfaces into the air, preventing the city from cooling off. In Montréal, at night and in the early morning, especially in winter, the city core is sometimes 5 to 8°C warmer than the environs. Daytime temperatures in summer reveal much smaller differences of 1 to 2°C between urban and rural settings.

Cities are, in fact, islands of heat within the surrounding countryside. A patchwork of warm zones over industrial areas, business centres, or commercial strips may coalesce to form the centre of the heat island. The boundary of the island is marked by steep temperature changes at the edge of the continuous built-up area. In general, as a city grows in size, the heat island expands and intensifies.

Above most cities is found a smoky, hazy canopy of polluted air -- the urban dome. Under light winds or calms the dome may extend upward for 1 1/2km, with the warmth and pollution mainly confined within the boundaries of the city. With a moderate breeze, the dome changes into a plume that can stretch downwind for at least 30 km before being dispersed.

As for cities, each wall, roof, or paving stone alters somewhat the climate above and around its vicinity. Where there are clusters of buildings, extensive asphalt and concrete surfaces, and a concentration of people in small areas, local and regional climates are greatly changed.

In cities we find not only different surface materials and surface shapes than in the country, but also different sources of heat, moisture, and atmospheric pollution. All of these differences have climatic consequences. In general, radiation, wind speed, visibility, sunshine, and heating requirements are less in urban areas than in nearby rural settings. The reverse is likely for temperature, cloudiness, thunderstorm frequency, and air pollution. For precipitation, however, the differences are not marked.

The Edmonton Municipal and Edmonton International Airport sites provide an excellent example of the effect of urbanization on climate. The International Airport is located about 20 km south of the city in flat farmland; the Municipal Airport lies within the city limits but about 3 km northwest of the main business district. The accompanying table summarizes the differences in several major climate elements for the two sites. These differences are averages of values collected over a long period of time.

The relatively greater warmth of cities has some advantages in cold climates. Wind chill, snowfall, and snowcover are reduced, and, on the economic side, snow clearance costs should be less. Heating requirements in buildings may also be reduced by as much as 13%. In summer, cities benefit from a longer growing season and a longer frost-free period but consume additional energy dollars through a greater need for air-conditioning. As cities expand, their effect on local climate becomes more pronounced.

URBAN-RURAL CLIMATE DIFFERENCES AT EDMONTON

Edmonton Municipal Airport	Element	Edmonton International Airport
	Temperature	
+2.8°C	January average minimum	
+2.6°C	July average minimum	
+1.5°C	Annual mean	
	Annual heating degree-days	+507(+8.5%)
+39	Annual cooling degree-days	
	Number of days <-20°C	+15
	Number of days <-2°C	+26
	Sunshine	
	Annual duration	+51 h(+2.2%)
	Precipitation (annual)	
	Rain days	+4
	Snow days	+2
	Thunder days	+5
	Number of January Observations with	
same	freezing rain/drizzle	same
	blowing snow	+6
+49	fog	
+20	smoke/haze	
+7	visibility <1 km	
	Wind	
	mean January speed	+0.4 km/h
+2.4 km/h	mean July speed	
	percentage of calms	+2.8%

WILF KENYON

Climate Elements

The climate of a place consists of a large number of weather components or elements, of which temperature and precipitation are the most commonly observed. Although most of us think about climate simply in terms of degrees of heat and cold and amounts of wet and dry, temperature and precipitation do not represent the totality of climate. Wind speed and direction, humidity, bright sunshine, cloud type and amount, evaporation, lightning, visibility, air pressure and soil moisture are also part of a place's climate. Indeed, there are so many climate elements that it is safe to conclude that no two places on earth have exactly the same climate.

This section describes some of the more common climate elements. The most comprehensive discussion is devoted to temperature and precipitation and their relations: frost, degree-days, snowcover and acid precipitation, among others. Two-page spreads are reserved for each of sunshine and radiation, wind speed and direction, and humidity and evaporation. In addition, considerable attention is given to drought, floods, sea ice and winter and summer storms.

The normal behaviour of the climate elements is described, their broad patterns from place to place are presented and their variations over time are explained on a national scale - setting the scene in the next section for a detailed treatment of the climates of the provinces and the territories.

Not all elements in Canada can be described with the same degree of detail. Temperature and precipitation are regularly measured at thousands of sites across the country. Some of them have histories that go back a hundred years or more. On the other hand, acid rain stations number 22 and only date back to 1973.

In 1989, the Canadian federal climate archive contained over three billion observations. In that year, measurements were received from 3131 observing sites. Twice-daily measurements of temperature and precipitation are the most commonly taken observations at 1873 sites. Hourly observations are made at 310 sites. Other climate stations in the federal network, mostly operated by corporations, government agencies, universities and utilities, have programs to measure solar radiation, evaporation, soil temperature and air quality.

Element	Number of Stations in 1990	First Observation in the Archive
Temperature	2,181	1840
Precipitation	2,537	1840
Upper Air Temperature	32	1961
Wind	498	1933
Sunshine	292	1951
Evaporation	126	1962
Soil Temperature	62	1958
Ozone	6	1960
Visibility/Ceiling	310	1947

TEMPERATURE

When we tune in the morning weather report, the day's temperatures are probably the first thing we want to know. Temperature, after all, affects us in many ways. It influences what we wear, how we feel and behave, what we plant and when, how much we spend on heating and cooling our homes, and where we go on our vacations. Temperature also affects other elements of the weather and the environment, determining whether it will rain or snow and controlling ground conditions, such as permafrost and ice cover.

Many people think of Canada as a cold country, but, in fact, most of the country enjoys a temperate climate. Although winters are indeed cold, summers are correspondingly hot. Temperatures can also shift dramatically from day to day. It is this range and variability of temperature that gives the Canadian climate much of its character.

What Affects Temperature?

Temperature varies most obviously with season and latitude. But the constant redistribution of heat by the earth's atmosphere also has an effect, and masses of air from the frozen north or the tropical south frequently bring extreme temperatures to many parts of the country.

Temperatures can sometimes vary considerably within a small area as a result of local factors, such as differences in altitude, the slope of the land, the proximity to water, urban heat effects, and the nature of the ground or surface cover.

On the average, temperatures near the ground will decrease at the rate of 6.5°C for every 1000 m rise of altitude. Western slopes are usually cooler in the morning and warmer in the afternoon than eastern slopes. Sandy soil will heat and cool more rapidly than a loam or clay soil or soil covered with vegetation. Bodies of open water moderate temperatures, making them cooler by day and warmer by night. Cities invariably average a few degrees warmer than the nearby countryside.

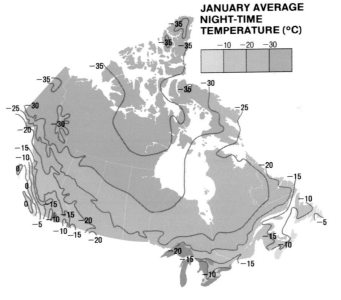

JANUARY AVERAGE NIGHT-TIME TEMPERATURE (°C)

−10 −20 −30

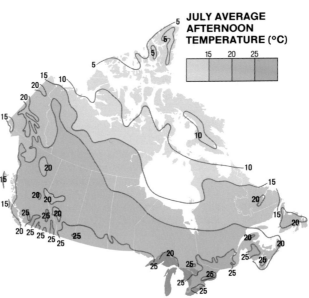

JULY AVERAGE AFTERNOON TEMPERATURE (°C)

15 20 25

Winter

The common image of the Canadian winter is one of universal and unrelenting frigidity, but, in reality, the amount of bitterly cold weather varies considerably from one part of the country to another. Climatologists define a very cold day as one on which the minimum temperature is below -20°C. In a single year, the number of very cold days in Canada ranges from near zero along the Pacific coast, Lake Erie, and the outer Atlantic coast to 210 days at Eureka in the Northwest Territories. In a typical year we can expect 5 very cold days in the B.C. interior, 60 across the eastern Prairies, 20 in central Ontario, 30 in the Gaspé, and over 100 in Yukon.

Inland temperatures are usually significantly colder than coastal temperatures. For example, in January (which is normally the coldest month of the year in Canada), the average temperature is a mild 2.5°C at Vancouver and a slightly frosty -3.9°C at St. John's but a chilling -19.3°C at Winnipeg.

The West Coast enjoys the shortest and mildest winters because of the moderating effect of warm Pacific air. Occasionally, however, the mainland coast and even the islands will experience a bout of more extreme weather when Arctic air spills over the mountains. Along the Atlantic seaboard the moderating effect of ocean air is less pronounced, since the prevailing air flow is off the land. Occasional influxes of warmer Atlantic air, however, maintain winter average temperatures along the Atlantic coast at about -2.5°C. Away from the coasts and mountains, monthly temperatures vary fairly directly with latitude. Regina's January average, for instance, is 10° higher than Tuktoyaktuk's on the Beaufort Sea.

In mountainous country, atmospheric inversions will often keep the mountain slopes warmer than the land below. At Faro, Yukon, for example, the January mean temperature is -24.5°C compared with -19.8°C at nearby Anvil, which is 464 m higher in elevation. Some of the coldest temperatures in Canada are found in the mountain valleys of Yukon and of Ellesmere Island.

Summer

Canadian summers can be hot. In mid-summer, the afternoon temperatures usually climb past the 25°C mark over much of the interior of the country. In coastal areas, however, sea breezes keep maximum temperatures 5 to 8° cooler than those at inland localities.

Summers are warmest in the dry interior B. C. valleys and in extreme southwestern Ontario. In these regions, the average daily afternoon temperature exceeds 27°C. The coolest, apart from the Arctic, are the western mountains and the Labrador coast, where daily highs in the summer average between 15 and 20°C, and daily lows between 5 and 10°C.

In the Arctic, there is a noticeable decrease in temperature as we move northwards, with average daily maximum temperatures declining from 10°C at the Arctic coast to 5°C at the northern end of Ellesmere Island. If we compare overall temperature averages for the summer, there is a 15°C difference between the northern tip of Ellesmere and the southern boundary of the Northwest Territories. In winter, the difference between these two latitudes is only half that amount.

Significant local cooling is often provided by the cold waters of the large inland lakes,

and of Hudson Bay where ice floes still drift in July. Shoreline localities around the bay or on the Great Lakes can experience a sudden change in temperature with a shift of the wind.

Climatologists define a *hot day* as one on which the maximum temperature is above 30°C. In Canada, almost every location south of the Arctic Circle experiences at least one hot day a year. There are 30 hot days in B.C.'s interior dry belt, 10 to 20 across the southern Prairies, 5 to 15 in southern portions of Ontario and Quebec, and less than 5 in the Atlantic Provinces.

Temperature Ranges

The difference between the lowest and the highest temperatures on a day is known as the *daily* range. The lowest temperature usually occurs around sunrise, the highest in mid-afternoon. The daily range is greatest in the summer, running from 10 to 14°C inland to less than 9°C in coastal areas. In winter, the range diminishes everywhere, with differences between the maximum and minimum temperatures averaging between 5 and 10°C.

The average *annual* range is obtained by subtracting the average temperature of the coldest month from that of the warmest month. It varies from 40°C in the central Northwest Territories and northern Manitoba to 10°C along the Pacific coast. Typical ranges are 35°C for the Prairies, 28°C for southern Ontario, 32°C for southern

and 90°C in the Prairie Provinces.

The Warmest and the Coldest

Canadian temperature extremes are not impressive when compared with world extremes. The lowest temperature observed in Canada was -63°C at Snag, Yukon, on February 3, 1947 — a North American record. The world record low temperature is -89.6°C, recorded at Vostok, Antarctica, on July 21, 1983. The highest temperature ever recorded in Canada was 45°C at Midale and Yellow Grass, Saskatchewan, on July 5, 1937. The world record is 58°C set at Al'azizyah, Libya, on September 13, 1922.

Temperatures lower than -40°C have been experienced in all regions except Prince Edward Island and higher than 40°C in all but the Maritimes and the Arctic.

Hot Spells and Cold Spells

Day-to-day weather change is a characteristic of the Canadian climate. On occasion, however, one type of weather can get locked in for a long time, creating social disruption and inconvenience. If the spell is severe and persistent, economic hardship and loss of life may result.

Every region has one or two periods a year with unusually wet or dry or cold or hot weather. Cold spells are relatively common on the Prairies. On occasion they intensify and persist over a large area for a few weeks or even months. During the

LOCAL TEMPERATURE VARIATION (°C)

	Average Midwinter		Average Midsummer		All-time Extremes	
	Max.	Min.	Max.	Min.	Max.	Min.
Altitude Differences						
Old Glory Mountain, B.C.	-8.4	-13.3	13.2	5.6	27.2	-37.8
Warfield, B.C.	-1.3	-6.0	28.3	13.3	41.1	-31.1

Old Glory at 2347 m; Warfield at 606 m; Stations are about 10 km apart.

	Average Midwinter		Average Midsummer		All-time Extremes	
City/Suburban Differences	Max.	Min.	Max.	Min.	Max.	Min.
Edmonton Municipal A	-10.7	-19.2	23.0	11.8	34.4	-44.4
Edmonton Int. A	-10.9	-22.0	22.4	9.2	35.0	-48.3

	Average Midwinter		Average Midsummer		All-time Extremes	
Lake/Land Differences	Max.	Min.	Max.	Min.	Max.	Min.
Caribou Island, Ont.	2.9E	-14.9E	14.5	11.6	25.6	-30.0E
Montreal River, Ont.	-8.4	-22.2	25.1	8.9	38.9	-47.7

E = Estimated; Stations at same latitude.

Quebec, 25°C for the Maritimes, and 20°C for Newfoundland.

Subtracting all-time extreme high and low temperatures reveals some incredible differences. The greatest absolute range is 100.5°C, at Fort Vermilion, Alberta. Generally, the extreme range is around 70°C in the Maritimes, 80°C in Ontario and Quebec,

cold spell of 1936, minimum temperatures each night at Saskatoon dipped below -18°C for 58 consecutive days from January 3 until March 1. Another memorable cold spell lasted 26 consecutive days when the temperature at Edmonton stayed below -18°C from January 7 until February 2, 1969. Eastern Canada usually experiences lengthy deep freezes each winter, but they are seldom as long or severe as in the West. The longest recent cold snap on record at Toronto was 7 consecutive nights with temperatures of -18°C or lower, beginning on February 8, 1979. Montréal's longest in recent years lasted 13 nights, beginning on January 5, 1968.

Heat waves occur commonly in southern Ontario and the interior of southern British Columbia. The rest of the country, however, rarely experiences more than three consecutive days with temperatures of 32°C or more at any one location, and no region ever experiences hot and humid spells as enervating as those common to the southern United States.

Canada's worst heat wave struck the western and central regions of the country in July 1936. For a week and a half, 32°C temperatures prevailed from southern Saskatchewan to the Ottawa Valley. High humidity added to the discomfort. Crops were blackened and 780 people died as a result of the extreme heat. Temperatures of 44.4°C at St. Albans, Manitoba, and 42.2°C at Atikokan, Ontario, recorded on the 11th, were all-time records for the respective provinces.

Heat waves in summer are much shorter than cold waves in winter. Most hot, humid spells break within 5 or 6 days. The longest consecutive string of hot days in Toronto, with maximum temperature at or above 30°C, lasted 12 days, starting on August 24, 1953.

ENVIRONMENT CANADA

Degree-Days

The "degree-day" is one of the most practical weather statistics ever devised. So useful has the degree-day concept become that daily, weekly, monthly, and seasonal totals are used routinely by those in business, industry, and government and by individual home owners and farmers. The degree-day measures the difference between the average temperature for a day and some reference temperature. Various reference temperatures are used for different purposes.

Heating. The best known degree-day is the heating degree-day based on a reference temperature of 18°C. To calculate the number of heating degree-days for a day, we subtract the day's average temperature from 18°. A day with an average temperature of 5°, for example, would contain 13 heating degree-days.

The consumption of fuel in most homes and buildings is reasonably well correlated with mean outside air temperature below 18°C. For every additional heating degree-day, proportionally more fuel is needed to maintain comfortable indoor temperatures. A day with 80 heating degree-days would therefore require eight times as much fuel as a day with 10 heating degree-days. Consumption rates are also fairly constant; for example, fuel consumed for 80 degree-days is about the same whether the 80 was accumulated on only one or two days or over seven days.

The average annual heating degree-day total (HDD) is about 3000 over the southern tip of Vancouver Island and less than 4000 across the southern coast and interior valleys of British Columbia and in the southwestern corner of Ontario. Totals increase to 5000 across the Maritimes and southern Alberta, reach 6000 over the southeastern Prairies, and 7000 to 8000 over northern portions of the western and central provinces. HDDs are between 10 000 and 13 000 across northern Canada in an average year.

Cooling. Temperature departures above 18°C or cooling degree-days (CDD) are used in assessing air-conditioning needs. Annual CDDs in Canada peak at an average of 400 near Windsor but are less than 300 throughout southern and eastern Ontario and near Montréal. The area

south of Winnipeg has about 200 cooling degree-days per year, and Regina has about 135. Most of the rest of the country has less than 100 CDDs per year.

The summer of 1988 was the third warmest on record in southern Ontario, where weather records date back to 1840. Extra electricity requirements for air-conditioning resulted in a record consumption of electrical energy that generated occasional power outages. Cooling degree-days from May to August in the Windsor—Toronto region ranged 30 to 40% above normal.

Growing. In general, the average annual number of degree-days above 5°C corresponds closely to both the duration and amount of heat needed for favourable plant growth. The growing degree-day (GDD) base may vary, however, for different plant species and for the prediction of various farming activities such as seeding or harvesting. Most Canadian agricultural areas have in excess of 1250 GDDs for hardy grains and hay and 1750 for tender fruits and vegetables. The growing season is anywhere from 6 to 10 weeks longer than the frost-free period.

Frost

Frost occurs whenever air temperatures fall to 0°C or lower. Its occurrence is of vital concern to agriculture. The choice of growing tender crops, such as beans and tomatoes, or hardier plants, such as potatoes or cabbage, is largely influenced by the number of days that are expected to be free of frost during the growing season. The variability in the occurrence of the last frost in spring and the first frost in fall is important in many crucial farming operations, such as scheduling planting and harvesting. The frost hazard is primarily a local agricultural concern; however, the impact of a severe frost may disturb the entire economy, affecting not only the prices of farm products, but jobs and incomes as well.

Several factors are significant in explaining the occurrence of frost. Local topography, including dif-

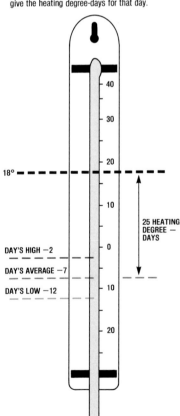

WHAT ARE HEATING DEGREE-DAYS?

Heating degree-days for a day in winter. The day's high temperature and low temperature are added and then divided by two to get an average temperature for the day. The average is subtracted from 18 to give the heating degree-days for that day.

18°

40
30
20
10
0
10
20

25 HEATING DEGREE — DAYS

DAY'S HIGH —2
DAY'S AVERAGE —7
DAY'S LOW —12

ferences of elevation, aspect, and slope may explain why some stations only a few kilometres apart and in apparently similar surroundings have frost-free periods that differ in length by as much as 45 to 60 days. Other factors affecting the incidence of frost are latitude, surface cover (vegetation, soil, snow), and proximity to water bodies and human settlements.

The frost-free season is the period normally free of sub-freezing temperatures, starting with the last frost in spring and ending with the first frost in the fall. It varies from over 240 days in parts of the Queen Charlotte Islands and Vancouver Island to less than 50 over the entire Arctic and the high mountains of western Canada. The northern forest zone counts 60 to 110 days, and across the Prairies there are between 80 and 120 frost-free days. Higher values, 120 to 140 days, occur in southwestern Canada. Around the Great Lakes, the frost-free season exceeds 160 days. Moreover, the cooling effect of the lakes retards the early development of fruits in spring when late frosts might strike. In the fall, the lakes exert a warming influence, which delays the occurrence of early frosts.

Thermal Comfort

Everyone knows that weather may cause discomfort. We know only too well that windy days with temperatures near freezing can feel just about as uncomfortable as calm days with temperatures several degrees lower. Prolonged windy cold may lead to frost bite, exposure, and hypothermia. Likewise, hot weather with high humidity feels uncomfortable and, if excessive and prolonged, may lead to heat-related cramps, exhaustion, or strokes.

Defining discomfort is difficult because personal reactions to the weather differ greatly according to a number of variables, including health, age, clothing, occupation, sex, and acclimatization. However, it is possible to come up with an approximate measure or index of discomfort that most people would consider reasonably accurate. Two climate discomfort indices used in Canada are wind chill and humidex.

Wind Chill...
IT'S NOT THE COLD IT'S THE WIND!

Wind chill is a simple measure of the chilling effect experienced by the human body when strong winds are combined with freezing temperatures. The wind chill temperature or heat loss factor is a good indicator of what one should wear to be protected from the cold. It does not, however, consider many other factors important for determining the body's heat retention or loss. Is the person sitting for hours as, for example, at a football game? Is the sun shining? What is the humidity, the insulating quality of the clothing, the age and the health of the individual?

Two means of expressing wind chill are heat loss in watts per square metre (W/m) and the equivalent temperature in °C. At 1400 W/m or -20°C, its wind chill temperature equivalent, weather is unpleasant and work or travel becomes hazardous unless heavy outer clothing is worn. The percentage of time that the wind chill exceeds 1400 or -20° in January is 83% at Winnipeg, 62% at Edmonton, 42% at Ottawa, 17% at Halifax, and under 1% at Vancouver and Victoria.

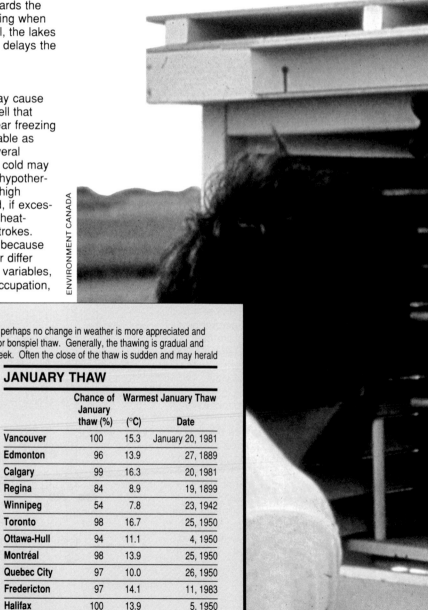

ENVIRONMENT CANADA

JANUARY THAW

For Canadians in the midst of a long winter perhaps no change in weather is more appreciated and better remembered than the January thaw or bonspiel thaw. Generally, the thawing is gradual and temporary, lasting from a few hours to a week. Often the close of the thaw is sudden and may herald the arrival of the coldest temperatures of the winter. For certain, there is still plenty of winter to come.

The cause of this upsurge in temperature is no mystery. The prevailing northwesterly circulation of Arctic air breaks down and mild westerly or southwesterly air spreads across Canada. What is a mystery is why this phenomenon so often occurs at about the same time year after year, usually during the third or fourth week of January. Its non-occurrence is very rare. At Toronto only one January since 1840 has not had a thaw, and at Edmonton only four Januarys in 100 have been thawless. The percentage chance of a January thaw for other major cities is: Montreal 98%, Halifax 100%, Winnipeg 54%, and Calgary 99%. So honoured is the January thaw in our weather lore that any mild "winterlude" is often referred to as the January thaw even when it occurs in December or February.

JANUARY THAW

	Chance of January thaw (%)	Warmest January Thaw	
		(°C)	Date
Vancouver	100	15.3	January 20, 1981
Edmonton	96	13.9	27, 1889
Calgary	99	16.3	20, 1981
Regina	84	8.9	19, 1899
Winnipeg	54	7.8	23, 1942
Toronto	98	16.7	25, 1950
Ottawa-Hull	94	11.1	4, 1950
Montréal	98	13.9	25, 1950
Quebec City	97	10.0	26, 1950
Fredericton	97	14.1	11, 1983
Halifax	100	13.9	5, 1950
Charlottetown	100	11.7	26, 1950
St. John's	100	15.2	9, 1979

Baker Lake, N.W.T., has the distinction of having the highest wind chill in Canada, an average of 1980 W/m² in January.

Humidex...

IT'S NOT THE HEAT, IT'S THE HUMIDITY

The expression most often heard in southern Ontario and southwestern Quebec on those hot, hazy, humid days of summer is "It's not the heat, it's the * !@#! ?#* ? humidity! " When high humidities interfere with the loss of heat from the body, because the air is already so moist it cannot take up all the moisture our bodies wish to evaporate, we are uncomfortable. In fact, we may actually feel warmer on a humid day than on a hotter but drier day. Humidity in association with air temperature, therefore, is a good measure of climatic discomfort. Several formulas have been developed to express this relationship between humidity and temperature on the one hand and discomfort on the other. Of these, humidex is the one most familiar to us in Canada.

Humidex is a measure of what hot weather "feels like" to the average person. It is analogous to wind chill in that it is an equivalent temperature index. Air of a given temperature and moisture content is equated in comfort to air of a higher temperature which has a negligible moisture content.

Humidex (°C)	Degree of Discomfort
20 - 29	Comfortable
30 - 39	Varying degrees of discomfort
40 - 45	Almost everyone uncomfortable
46+	Active physical exertion must be avoided
54+	Heat stroke may be imminent

Of course comfort is subjective and largely dependent on the age and state of health of the individual. Weather conditions causing heat cramps in a teenager may result in heat exhaustion for a middle-ager and heatstroke for seniors. On average, 11 Canadians die annually from excessive heat and sun-stroke. The number who suffer heart attacks and other ailments from hot weather discomfort is unknown. Prolonged heat stress is unusual in Canada outside southwestern Ontario and on occasion in southeastern Manitoba and southwestern Quebec.

At a humidex value of 30°C, some people begin to experience discomfort. The percentage of time the humidex exceeds 30°C for one hour or more in the peak of summer is 71% at Windsor, 47% at Winnipeg, 45% at Ottawa, 24% at Charlottetown, 11% at Edmonton, and 4% at Vancouver. North of 55°N, uncomfortable humidex readings are very rare.

The highest values occur at Windsor, where at the peak humidex hour of 3 p.m., the mean humidex during the last week of July is 32°C. Extreme values above 48°C have occurred at several locations in southwestern Ontario.

HIGHEST HUMIDEX AND WIND CHILL VALUES (°C)

by Province / Territory

	Humidex	Wind Chill
British Columbia	53	-69
Alberta	47	-68
Saskatchewan	52	-70
Manitoba	54	-76
Ontario	52	-70
Québec	50	-77
Nova Scotia	45	-53
New Brunswick	49	-61
Prince Edward Island	41	-57
Newfoundland	46	-71
Northwest Territories	47	-92
Yukon	40	-83

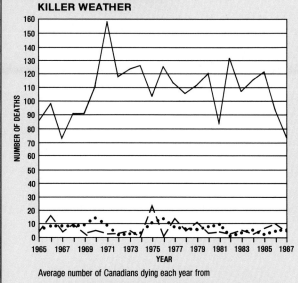

KILLER WEATHER

Average number of Canadians dying each year from
—— Excessive cold (average 110)   — — Excessive heat (average 7)
•••• Lightning (average 7)

KILLER COLD

Exposure to extreme cold claims more lives in Canada than lightning, storms, floods, heat waves, tornadoes, and any other weather phenomenon. Mortality figures, compiled by Statistics Canada over a 15-year period, show a total of 1,621 deaths attributed to "excessive cold." This amounts to an average of 108 deaths per year or about 0.5 deaths per 100,000 persons. Cataclysms — a composite category including deaths caused by storms, floods, earthquakes, tidal waves, and other natural events — claim only 17 Canadians a year on average.

Climate Severity

The climate severity index describes in a single number many of the unfavourable (uncomfortable, depressing, confining, and hazardous) aspects of the Canadian climate. The extreme values of 18 climate parameters, including wind chill, humidex, length of winter and summer, wet days and fog days, and strong winds are combined into an index that ranges from 0 to 100. The index has proven quite popular as a guide for those selecting the more amenable climates for recreational and retirement living, for those employers concerned with the timing of outdoor activities and performance, and for workers seeking fair and equitable remuneration for working outdoors.

Four major factors make up the index, namely: discomfort, psychological state, safety, and outdoor mobility. Each factor is a combination of several climate elements. To produce a severity index, points are assigned to each of the elements on the basis of their intensity, duration and other such criteria. The points are then totalled and adjusted mathematically to give the final score.

Considering the importance of comfort in our daily lives, the discomfort factor, separate for winter and summer, is treated as the most important one and accounts for more than half of the severity index. Three elements in defining winter discomfort are wind chill and the duration and severity of winter. Summer discomfort is defined by considering humidex, length and warmth of summer, and dampness.

The other three factors are each given equal weight and judged to be of lesser importance than discomfort because they are generally associated with less frequent and more ephemeral conditions. The climate elements that best represent psychological state are length of the winter day, lack of bright sunshine, wet days, and fog frequency. The general hazardousness of a locality can be described by considering snowfall and the frequency of strong winds, thunderstorms, and blowing snow. Outdoor mobility can be measured by assessing snowfall, freezing precipitation, and poor visibility.

The climate severity index is designed so that values approaching 100 indicate the highest severity. In Canada, much of the northern Queen Elizabeth Islands, the Beaufort Sea coast, and the Hudson Bay coast have the highest severity, with all four factors showing high values. Yukon, the northern Prairies, northern Ontario, most of Quebec, and Newfoundland also have high severity indices. Low severity scores occur over southern British Columbia, southeastern Alberta, and southern Ontario. The least climatic severity in Canada is found along the east coast of Vancouver Island, the lower Fraser Valley, and the southern interior valleys of British Columbia.

The accompanying table shows index values for twelve large Canadian cites.

ENVIRONMENT CANADA

SEVERITY INDICES FOR METROPOLITAN AREAS ARRANGED IN DECREASING ORDER OF POPULATION	Climate Severity Index
Toronto	35
Montréal	44
Vancouver	18
Ottawa-Hull	43
Edmonton	37
Calgary	34
Winnipeg	51
Québec City	52
Victoria	13
Regina	47
St. John's	56
Saint John	48

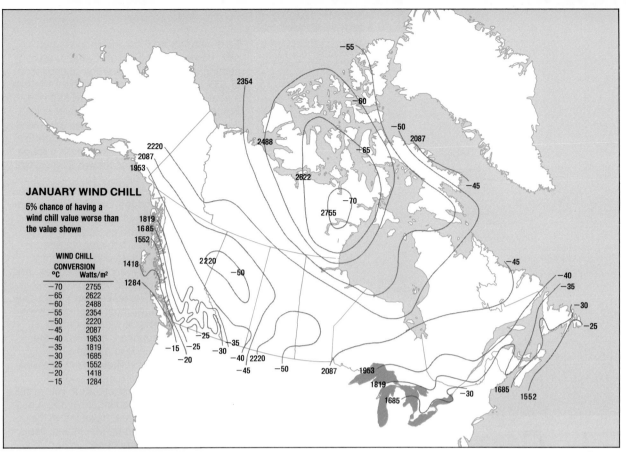

JANUARY WIND CHILL

5% chance of having a
wind chill value worse than
the value shown

WIND CHILL CONVERSION	
°C	Watts/m²
−70	2755
−65	2622
−60	2488
−55	2354
−50	2220
−45	2087
−40	1953
−35	1819
−30	1685
−25	1552
−20	1418
−15	1284

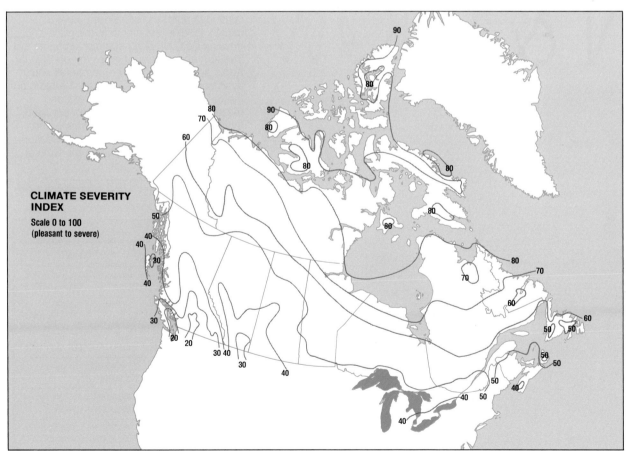

CLIMATE SEVERITY INDEX

Scale 0 to 100
(pleasant to severe)

RAIN & SNOW

Each year about 5 1/2 trillion tonnes of rain, snow, and hail fall on Canada, representing a nation-wide average precipitation of 535 mm.* For comparison, the earth has a yearly precipitation average of 690 mm. Somewhat more than 60% of this precipitation eventually runs off into lakes and rivers. The remainder evaporates from the earth's surface or passes back into the air through plant tissues (a process known as transpiration).

The oceans, principally the Pacific, the Gulf of Mexico, and the Caribbean Sea, are the primary sources of water for precipitation in Canada. However, water may recycle several times between air and ground while moving across the country, evaporating from soil and water bodies, and transpiring from vegetation.

About 36% of Canada's yearly precipitation occurs as snow, most of which covers the ground for several months before melting. By comparison, only 5% of the earth's precipitation falls as snow.

Precipitation is distributed very unevenly across Canada in both space and time. The Pacific Coast is the wettest area, with most of its rain and snow coming in winter. Less than 100 km to the east, however, in the rain-shadow of the Coast Mountains, lies a sagebrush-dotted semi-desert, one of the driest areas in the country. The rain-shadow continues across the Prairies. June is usually the wettest month in the grain-growing belt, although summer droughts are frequent and severe. Across the north is the Arctic desert, with its meagre rainfall and even scantier snowfall. Eastern Canada, on the other hand, usually has ample and reliable quantities of precipitation provided by moist air from the south. Ontario and Quebec have no special wet or dry season, whereas the end of the year is the wettest time for the Atlantic coast.

*Local averages range from 64 mm at Eureka, N. W. T., to 6655 mm at Henderson Lake, B. C.

PRECIPITATION RECORDS

	Canada	World
greatest precipitation in 24 hours	489.2 mm Ucluelet Brynnor Mines, B.C. Oct. 6, 1967	1869.9 mm Cilaos, La Réunion. Island March 15, 1952.
greatest precipitation in one month	2235.5 mm Swanson Bay, B.C. Nov. 1917	9300 mm Cherrapunji,. India July 1861.
greatest precipitation in one year	8122.6 mm Henderson Lake, B.C. 1931	26 461.2 mm Cherrapunji,. India Aug 1860 - July 1861.
greatest average annual precipitation	6655 mm Henderson Lake, B.C.	11 684 mm Mt. Waialeale,. Kauai, Hawaii.
greatest average annual snowfall **	1433 cm Glacier Mount Fidelity, B.C.	1460.8 cm Rainer Paradise. Ranger Station, Wash.
greatest snowfall in one season	2446.5 cm Revelstoke/Mount Copeland, B.C. 1971-72	2850 cm Rainer Paradise. Ranger Station, Wash.1971-72
greatest snowfall in one month	535.9 cm Haines Apps No. 2, B.C. Dec. 1959	990.6 cm Tamarack, Cal. Jan. 1911.
greatest snowfall in one day	118.1 cm Lakelse Lake, B.C. Jan. 17, 1974	193.0 cm Silver Lake,. Colo. April 14-15, 1921.

** Only Canada and the United States measure snow depth. Other countries measure snow amount in terms of the water equivalent.

Precipitation Totals

WHO GETS HOW MUCH?

The heavy precipitation that falls on the outer Pacific coast is the result of moisture-laden winds encountering the Coast Mountains of British Columbia and the St. Elias Mountains of southwestern Yukon. Here the average rain and snow total is as much as 3200 mm. In sharp contrast, annual amounts in the rain-shadow of the Coast Mountains average only 400 mm, with some localities receiving less than 250 mm. Farther east, the Prairies lie in the rain-shadow of the entire Cordillera. Annual precipitation totals in this dry belt average 350 to 500 mm. Locally, totals are 10 to 20% more over higher terrain. Additional moisture, originating over the Gulf of Mexico, enters Manitoba from the south, giving it an average 100 mm more precipitation each year than the western Prairies.

Moving eastward, precipitation increases at the rate of about 40 mm every 100 km, going from 500 mm a year at Winnipeg to 1500 mm at Halifax. Over 1000 mm of precipitation fall in the elevated areas to the lee of the Great Lakes, over Quebec's Laurentian Uplands, and over the Atlantic Provinces. The Appalachian Highlands and the Atlantic coast receive an additional 400 to 600 mm a year.

Across the north, melting permafrost, extensive snowcover, ice-caps, and thousands of lakes suggest plentiful precipitation. In fact, the area averages a meagre 100 to 200 mm per year, making it the driest region in Canada. Totals exceed 500 mm, however, over Baffin Island's east coast mountains.

The Most and the Least

All high precipitation records in Canada for rain or snow and for periods from hours to years are held by stations in British Columbia. No world records for precipitation exist in Canada. The accompanying table provides a comparison of Canadian and world precipitation records.

At the other extreme, the driest station on average in Canada is Eureka, N.W.T., with 64 mm a year. (This value may be slightly less than the true amount, however, since frequent exposure to strong winds causes rain and snow gauges to under-record the actual accumulations.) Rea Point, N.W.T., holds the Canadian record for the longest period without measurable precipitation: 218 days. In southern Canada, Ashcroft, B.C. is the driest station, with 236 mm of precipitation in an average year, and Frontier, Saskatchewan, has the longest dry period, 158 days.

Other parts of the world have more extreme records for low precipitation. The great deserts of the world sometimes go

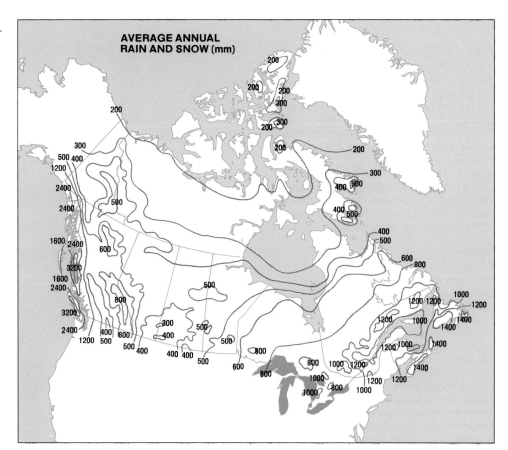

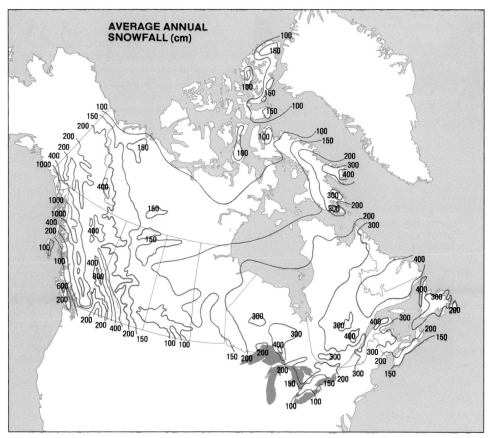

without any rain for several years in a row.

Days of Rain and Snow

Measurable precipitation (0.2 mm or more) occurs on an average of more than 100 days a year over nearly all of Canada. In the wettest region of British Columbia and in Newfoundland more than 200 days are wet. In the dry areas of western Canada, where annual precipitation is under 400 mm, precipitation days count fewer than 100 a year. In the east, the number of wet days varies between 125 and 150. The eastern side of Baffin Island has about the same number, but over most of the Arctic only 75 to 100 days have measurable precipitation.

Snow

PROBLEMS AND PLEASURES

For many Canadians snow is a nuisance and a hazard. Heavy loads of snow collapse buildings, blizzards and avalanches create perils, and quick thaws bring floods.

Overcoming snow costs our economy millions in snow shovels, snowploughs, and snow tires. Fortunately, we spend even more on having fun in the snow. Snow means snowballs, snowmen, and snow skiing. Snow transforms fall decay into winter glistening. Farmers appreciate snow's insulating quality and its moisture quantity, and wildlife depend on it

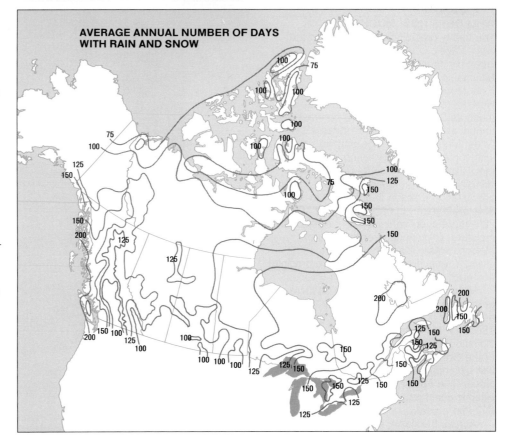

AVERAGE ANNUAL NUMBER OF DAYS WITH RAIN AND SNOW

GLORIA STEWART

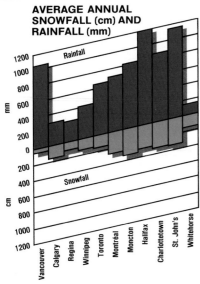

AVERAGE ANNUAL SNOWFALL (cm) AND RAINFALL (mm)

for shelter and survival.

Snowfall is difficult to measure. Before it can be counted it may drift, blow, evaporate, or melt. Annual snowfalls in any one location vary widely. More snow may fall in one day than fell in the entire previous winter.

The map of average annual snowfall shows a general pattern of lower amounts in the north and in the interior, and higher amounts in the east and in all elevated areas. Nowhere is the pattern more complex than in British Columbia, where you find Canada's highest and lowest snowfalls. The southernmost tip of Vancouver Island seldom receives snow, and what it does receive rarely lasts. An exception occurred in December 1968, when Victoria had a snow cover of 52 cm.

When Pacific air streams mount the higher parts of the Western Cordillera, i.e., the Coast and Rocky Mountains of British Columbia and the St. Elias Mountains of Yukon, record snowfalls of more than 1000 cm a year may occur. Near Revelstoke, 2446 cm of snow fell during the winter of 1971-72.

Large amounts also occur over Quebec's Laurentian slopes, over the Appalachian Highlands, and over the

Cumberland and Hall Peninsulas of Baffin Island. Newfoundland is one of the snowiest provinces, with several areas receiving 400 cm or more a year. There are also the great snowbelts in the lee of large open lakes and Hudson Bay. Cold air crossing open water gobbles up huge amounts of moisture, only to release it in snow bursts across the terrain on the lee side.

Winter rain, not snow, is likely in southwestern coastal British Columbia, over southernmost Ontario, and over the extreme southern and eastern coasts of Nova Scotia. The Prairies are also a low snowfall region, with many places having half the snowfall amounts experienced in eastern Canada. It comes as a surprise to many southerners to find that the Arctic has one of Canada's lowest snowfalls, 50 to 100 cm a year, despite having some snowfall in every

DOUGLAS McCOLLOR

32

month. This lack of abundant precipitation reflects the weakness of passing storm systems and the dryness of the atmosphere.

Climatologists count days with 0.2 cm or more of snow as snow days. Snow days in British Columbia range from 10 to 20 along the Pacific coast, 20 to 50 in the southern valleys and 75 and more in the northern portion and elevated areas. Values in Yukon range from 50 along the Arctic coast and over most of the Territory to 100 in Kluane National Park. The Prairies have between 35 and 50 snow days and the Great Lakes region around 50, though Windsor has 20 days and the snowbelts have 75 days a year. Across southern Quebec and Newfoundland, there are 50 to 75 snow days a year but less than 50 over most of the Maritimes.

Paul Bunyon Snowfalls

One of Canada's more memorable snowfalls occurred in 1963 when 112 cm of snow fell over a 24-hour period at Livingstone Ranger Station in southwestern Alberta. This stood for 11 years as a record for the greatest 1-day snowfall in Canada. What was even more remarkable about this event, however, was the date - June 29, right into summer. The current record for the greatest snowfall in one day belongs to Lakelse Lake, B.C., with 118.1 cm on January 17, 1974.

The record for the greatest snowfall in one month is held by Haines Apps, B.C., where 535.9 cm fell in December 1959. The greatest depth of accumulated snow on the ground is 483 cm, measured on March 31, 1974, at Whistler, Roundhouse, B.C., a pack deep enough to bury a one-storey building. Finally, the greatest snowfall over five consecutive days belongs to Pointe-des-Monts, Quebec, which received 248.9 cm in March 1885, and Kitimat, B.C., with 246 cm in January 1974. Glacier Mount Fidelity, B.C., has the greatest average annual snowfall in Canada: 1433 cm.

The Snow Season

Canada is snow country for a good part of the year. Some snow melts or evaporates on contact. More often it stays, compacts, and is added to by fresh snow, building a cover which lasts for several days or months. The duration of snowcover varies considerably from year to year. In some high elevations and high latitudes it never leaves but gradually accumulates and grows into glaciers and ice-caps.

Annual snowcover is important for protecting the root systems of perennial and winter crops from cold damage and small wildlife from the rigours of winter. It also serves as a major source of water for power, irrigation, and domestic supply.

The beginning of the snow season varies from mid-September across the southern Arctic islands to the end of December on the east shores of Newfoundland and Nova Scotia and along the outer coast of British Columbia. Typical dates for the formation of a lasting snowcover are the end of October for the northern Prairies, mid-November for

"PAUL BUNYON SNOWFALLS" GREATEST SNOWFALL OVER FIVE CONSECUTIVE DAYS

	Snowfall (cm)	Station	Starting Date
British Columbia	246.2	Kitimat	January 14, 1974
Alberta	169.2	Columbia Icefield	November 15, 1946
Saskatchewan	97.0	Cypress Hills	May 28, 1982
Manitoba	121.9	Morden	March 5, 1916
Ontario	197.6	Nolalu	December 4, 1977
Quebec	248.9	Pointe-des-Monts	March 16, 1885
Nova Scotia	121.9	Whitehead	March 3, 1916
New Brunswick	124.8	Turtle Creek	December 24, 1970
Prince Edward Island	92.7	Charlottetown	December 30, 1921
Newfoundland	182.0	Cartwright	January 1, 1965
Northwest Territories	153.7	Fort Resolution	November 19, 1947
Yukon	86.4	Carcross	February 18, 1923

METRO TORONTO REGIONAL CONSERVATION AUTHORITY

There's No Business
LIKE SNOW BUSINESS

❄ In 1986, $2,807,000 worth of snowshoes were manufactured in Canada

❄ Canadians spend more on skiing than on snow removal; in 1987, the number of Canadian households with downhill skies was 1,774,000, with cross-country skies 2,568,000 and with snowmobiles 645,000.

❄ The number of ski resorts operating in Canada in 1985 was: 206 with an earned revenue of $158,000,000.

❄ In 1988, 678,310 pairs of skies were imported into Canada, worth $32,059,000.

❄ The snow removal budget for Montréal (1987-88) was $47 million; for Victoria (1985-87) $ 2,000.

❄ Montréal removes more snow from its streets than any other large city in the world — 42 million tonnes of snow in an average year.

❄ More than 10% of the operating budget of a ski area may be spent on snowmaking.

❄ In Ottawa 15% of the taxes are spent on snow removal; and snow removal is the fifth most costly city program (more than a quarter of the police budget and half that of fire protection).

❄ In 1975, the Armed Forces spent $200,000 in British Columbia releasing avalanches.

the southern Prairies and Quebec's north shore, and early December for southern Ontario.

The end of the snow season varies from mid-January on the outer west coast to the end of June in the Arctic Archipelago. Across the Prairies, snow disappears during early April in the south and late April in the north. Southern Ontario is snow-free by the first week in April, and Newfoundland by the end of April. Across the central Northwest Territories snowcover is usually scanty by June 1.

Acid Rain and Snow

Acid rain is acknowledged to be one of the greatest environmental threats of our time. No longer a local phenomenon, acid rain has become a hemispheric problem. By some accounts, affected regions have grown in size over time and the acidity of precipitation may have increased as well.

The acid rain story begins in the smoke-stacks of factories and the exhaust pipes of cars. Sulphur and nitrogen oxides are emitted into the atmosphere. Winds and weather carry these pollutants hundreds to thousands of kilometres away, allowing time for their transformation into compounds which combine with water vapour in the air to form mild sulphuric or nitric acids. They return to earth eventually in acidified rain or snow, in fog droplets, dew, and frost, or as dry matter. Without question, some lakes

and streams and their wildlife, forests, and soil are being adversely affected by acid rain. Also alarming is the accelerated decay of buildings and monuments and the possible danger that acid rain may have on human health.

For most of the winter, potent amounts of sulphuric acid accumulate with successive snowfalls and are held in the snowpack. Almost all of the acid is released suddenly

LONGEST WET AND DRY SPELLS IN DAYS WITH STARTING DATE

	Wet Spell		Dry Spell	
Toronto	16	Jan. 23, 1947	38	Feb. 3, 1968
Calgary	19	June 19, 1969	71	Aug. 31, 1885
Regina	15	Jan. 23, 1974	68	Oct. 4, 1888
Saskatoon	15	Dec. 18, 1968	88	Dec. 25, 1911
Thunder Bay	11	Sept. 27, 1983	34	Mar. 22, 1984
Winnipeg	12	Sept. 17, 1946	47	Oct. 9, 1976
Montréal	12	July 19, 1980	21	July 28, 1947
Québec City	18	Nov. 16, 1959	20	July 29, 1947
Saint John	13	Nov. 3, 1969	17	July 1, 1968
Fredericton	15	June 1, 1977	19	Mar. 4, 1962
St. John's	23	Feb. 18, 1970	14	July 3, 1952
Halifax	17	Jan. 2, 1982	18	July 1, 1982
Sydney	22	Dec. 6, 1962	18	Oct. 22, 1947
Charlottetown	22	Jan. 4, 1955	19	Jan. 21, 1961
Windsor	12	Nov. 21, 1979	31	Oct. 3, 1944
London	23	Jan. 19, 1947	30	June 14, 1963
Edmonton	10	May 21, 1980	79	Dec. 14, 1930
Victoria	33	Apr. 19, 1986	53	July 18, 1986
Vancouver	29	Jan. 6, 1953	62	July 6, 1928
Kamloops	15	May 25, 1980	63	Feb. 8, 1905

OBSERVING PRECIPITATION

Rain, drizzle, freezing rain, freezing drizzle, and hail can all be measured with a standard Canadian rain gauge. The gauge is usually placed on a level, grassy surface where it is free from the shelter of trees, fences, and buildings. The rim of the gauge is 40 cm above the ground and has a circular opening 11.3 cm in diameter. The rain is funnelled into a clear plastic cylinder which measures the contents to the nearest 0.2 mm.

Snowfall amounts are often calculated by using a standard ruler to measure the depths of freshly fallen snow at a number of representative points and recording the average to the nearest 0.2 cm. Snowdrifts and windswept bare spots are avoided as much as possible when making an observation.

At most ordinary climate stations, the water equivalent of the snowfall (in other words, the amount of rain it is equal to) is obtained simply by dividing the snowfall by 10. Principal climate stations, and some ordinary ones as well, are equipped with a shielded Nipher snow gauge, the contents of which are melted to obtain the water equivalent. The Nipher gauge has a horn-like shield that reduces the tendency of snow to blow across the top of the gauge instead of into it.

Total precipitation is recorded as rainfall plus the measured water equivalent of the snowfall, with snowfall being the depth of newly fallen snow. Since the water equivalent, and hence the depth of freshly fallen snow, varies with temperature and other factors, readings from snow gauges are more precise than those obtained from the snow ruler.

At most stations, precipitation measurements are made twice daily — in the morning, and in the late afternoon. At some stations only one observation is taken daily, whereas at principal stations observations are taken four times daily.

with the first meltwater in the spring runoff. As the snowmelt enters lakes and rivers, parts of these water bodies can become as much as 100 times more acidic than normal for a short time. Such a sudden and intense dose of acid produces a severe chemical shock on aquatic life.

Some Facts about Acid Rain

- Acid rain in eastern North America has a strong annual cycle with maximum acidity and sulphate concentrations occurring in late summer and autumn.
- Deposition is highly variable, with more than 50% of the annual deposition at eastern stations occurring in only 20% of the precipitation days.
- One of the worst single episodes of acid precipitation occurred between August 28 and September 4, 1981, when a heavy rainfall of 200 mm in the Muskoka-Haliburton region of Ontario accounted for a quarter of the region's total acid rain for that year.
- Geographical regions with high emissions of pollutant gases also tend to have high acidity in nearby water bodies.
- 50% of the sulphur deposited in Eastern Canada is from sources in the United States.
- No trend in the average annual pH of water bodies can be confirmed from our present intermittent, short-period records.

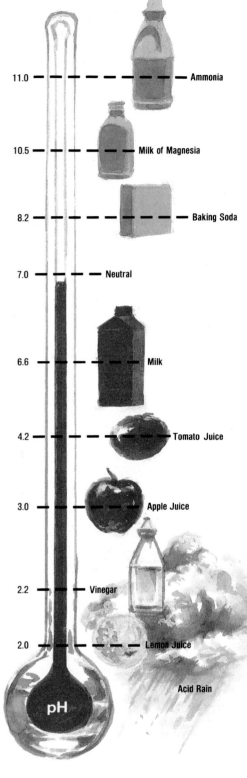

ACID RAIN

- ● Areas having SO₂ emissions greater than 100 kilotonnes per year
- Areas most sensitive to acid precipitation
- ● Favoured storm tracks

pH SCALE

A pH scale is used to measure whether a liquid is acidic or alkalinic. A pH of 7 indicates neutrality. The greater the acidity, the lower the pH value; the greater the alkalinity, the higher the pH value.

The scale is logarithmic so a drop in pH from say 4 to 3 corresponds to a 10-fold increase in the acidity of the liquid; and from 5 to 2, one thousand times more acidic.

Clean rain is slightly acidic with a pH of 5.6. The rain that falls over eastern North America now has a pH of 4.6 or lower — ten times as acidic as "normal" rain.

Sunshine Country

The sun shines over Canada every day of the year as it has for billions of years. Possible sunshine, the total number of hours of sunlight in the absence of obscuring clouds, mist, fog, dust, or terrain, depends only on latitude. At the 49th parallel Canada could receive 4479 hours of sunlight annually; at 60°N, the latitude along the top of the Prairies, 4536 hours, and at 80°N, 1000 km from the North Pole, 4567 hours. Cloud cover, especially in late fall and winter, pollution, and shadows reduce the amount of sunshine actually received to about half these durations or less in most seasons.

The sunniest places are beneath the high clear skies of the southern Prairies, where the average duration of bright sunshine exceeds 2400 hours a year. Only slightly less sunny is central British Colum-

bia, the remaining Prairies, and southern Ontario. Despite the high rainfall, most of the Maritimes have about 1800 sunshine hours a year. The lowest figures are along the Pacific coast just south of the Alaskan Panhandle, where the range is from 1200 to 1400 hours a year, and over fog-enshrouded parts of Newfound-

land, where fewer than 1500 hours of bright sunshine occur in an average year.

In the summer, most of southern Canada receives over 250 sunshine hours a month, about 60% of the possible amount, for an average of 8 1/4 hours a day. Again, the duration of sunshine in the summer months is greatest over the southern Prairies (more than 340 hours per month), and the lowest figures are along the coast of northern British Columbia — a dismal 120 hours or 4 hours a day on average.

Cloudy weather and the lack of sunshine are common across the country in winter but are most pronounced along the northwest Pacific coast. Prince Rupert, for example, records clear

THE METEOROLOGIST'S CRYSTAL BALL

Meteorologists really do use a crystal ball, or at least something that looks like one. This one, however, can't predict the future. It's the Campbell-Stokes recorder, and it's used for measuring periods of bright sunshine. The device is ingeniously simple. If you've ever used a magnifying glass to focus the sun's rays on a piece of paper and burn a hole in it, you will understand how it works.

The instrument is located away from buildings to prevent shading. By means of a glass sphere, the sun's rays are focused on specially prepared cards, leaving a charred track for those portions of the day when the sun is shining brightly. The charts are changed daily and the charring is measured and recorded in tenths of hours.

skies only 15% of the time in December, with an average of 32 hours of sunshine.

The seasons north of the Arctic Circle (67°N) are marked by periods of continuous winter darkness and summer light. At Cambridge Bay (69°N), there is continuous daylight from May 22 to near midnight on July 23, a period of 63 days. In winter, Cambridge Bay is in continuous darkness for 42 days, from shortly after noon on December 1 to January 13.

Radiating All Over This Land

As the sun's radiation passes through the earth's atmosphere, some of it is reflected back to space, some is absorbed, and some is scattered through the atmosphere. The solar energy which arrives at the earth's surface is a mixture of direct radiation and scattered — or diffuse — radiation.

We use the term *global radiation* to refer to the total solar radiation coming from the sun and all of the sky. This includes not only the direct beam from the unclouded sun but also the diffuse radiation from the scattered sunlight. The bluish-grey light of a cloudy day is mostly diffuse radiation. Even on a bright, sunny day, about 20% of solar radiation is diffuse. Bright sunshine, however, is the major component of global radiation, and those areas that receive the most direct sunshine also have the highest global radiation totals.

Global radiation values are lowest in areas where cloudiness is prevalent, such as along the coasts of the oceans and the lee shores of large inland bodies of water. The amount of obscuring smoke or haze in the air also influences radiation values.

During fall and winter, global solar radiation amounts vary with latitude, with higher values in the south and lower values in the north. Above the Arctic Circle, solar radiation values drop to zero during the continuous darkness of the period of polar night.

In spring, the largest amount of global radiation is received in central Canada between Lake Winnipeg and the Arctic coast. This area receives about 35% more radiation than the Atlantic or Pacific coasts. However, at this time of year, much of this incoming radiation has little effect in melting snow, evaporating moisture, or warming the ground. Most of the incoming energy — about 60% or more — is reflected off the bright snow cover.

In early summer, day length exceeds 15 hours everywhere in the country, and solar radiation is at its yearly high. There is relatively little variation in solar radiation totals across Canada during the summer months. Interestingly enough, the highest values occur in June over Ellesmere Island, the result of 24-hour daylight and clear skies. Here radiation totals are similar to values recorded in Texas and coastal California.

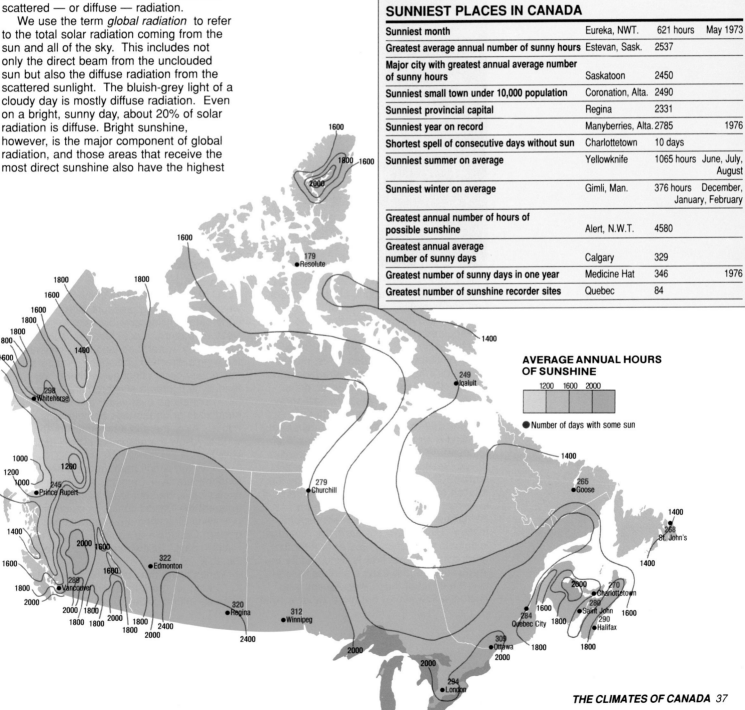

SUNNIEST PLACES IN CANADA

Sunniest month	Eureka, NWT.	621 hours	May 1973
Greatest average annual number of sunny hours	Estevan, Sask.	2537	
Major city with greatest annual average number of sunny hours	Saskatoon	2450	
Sunniest small town under 10,000 population	Coronation, Alta.	2490	
Sunniest provincial capital	Regina	2331	
Sunniest year on record	Manyberries, Alta.	2785	1976
Shortest spell of consecutive days without sun	Charlottetown	10 days	
Sunniest summer on average	Yellowknife	1065 hours	June, July, August
Sunniest winter on average	Gimli, Man.	376 hours	December, January, February
Greatest annual number of hours of possible sunshine	Alert, N.W.T.	4580	
Greatest annual average number of sunny days	Calgary	329	
Greatest number of sunny days in one year	Medicine Hat	346	1976
Greatest number of sunshine recorder sites	Quebec	84	

AVERAGE ANNUAL HOURS OF SUNSHINE

1200 1600 2000

● Number of days with some sun

SOMETHING IN THE *WIND*

LOIS TREMBLAY

Plainly, wind may be pleasant or unpleasant, depending on its strength, on the temperature and humidity, and on what that "something in the wind" is. For those beleaguered by summer's oppressive heat and humidity, any wind feels refreshing. Wind ventilates air-polluted cities, moves ice, propels sailboats, and generates power. On the other hand, wind chills air, drifts snow and soil, and drives rain, hail, spray, and waves. It also brings together air streams that may produce a thunderstorm.

Wind is air in motion. Differences in air temperature create differences in atmospheric pressure which, in turn, initiate air motion. This movement is modified by the rotation of the earth, which causes the wind to deflect to the right in the Northern Hemisphere. Another important factor is friction. High above the earth the flow of air is smooth and continuous. At or near the ground this uniform flow is disturbed by the irregularities of the earth's surface. Generally, the effect of these irregularities is to make the flow lighter but more variable, with lulls and gusts.

All of the provinces lie in the zone of the westerlies. Moving pressure cells, however, contribute considerable diversity to the westerly pattern. Local features, such as topography, water, and land use, influence the wind's direction and speed. Topography has a very great effect everywhere. Winds tend to be lighter in sheltered areas but much enhanced over water and on exposed headlands or on the tops of mountains or ridges. Also, there is a tendency for the wind to follow the alignment of well-defined valleys.

In winter, winds from the west and north predominate over Canada. A notable exception is the southeast winds that frequent the Pacific coast. In summer the prevailing directions are southwest in eastern Canada and southeast along the Pacific coast. Elsewhere, winds are quite variable.

Winds along the coast are stronger by 5 to 10 km/h or more than winds a short distance inland. The Gulf of St. Lawrence, the Atlantic and Pacific coasts, and Hudson Bay are typically the windiest locations in Canada. Cape Warwick on Resolution Island, N.W.T., is the station with the highest average annual wind speed: 35.3 km/h. At the other extreme, Dawson, Yukon, has an average speed of only 3.7 km/h over the year, with calm (no wind) conditions well over half the time.

Maximum winds are generally strongest in winter, but paradoxically the frequency of calm or light winds is also highest at this time of year, especially across the Prairies and the north. Generally, across Canada, the average wind speed is highest in spring. On the Prairies, spring winds average 18 to 20 km/h and, when loaded with snow or soil particles, can make travelling or farming difficult. Summer winds in Canada are generally light, less than 15 km/h, except in thunderstorms, frontal systems, or hurricanes. Wind speeds strengthen in the fall, with most locations having average values above 15 km/h.

Measuring Wind

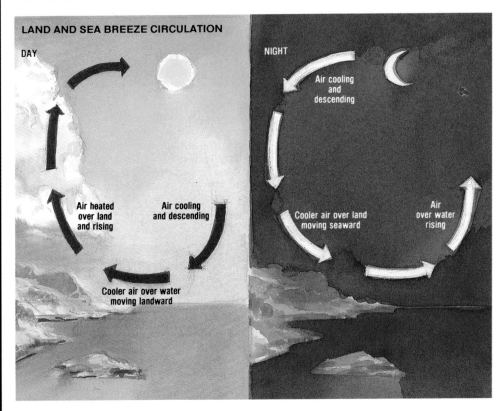

Environment Canada uses a vane to measure wind direction and an anemometer consisting of three conical cups to measure speed. The equipment is mounted on a tower at a height of 10 m above the ground and is situated over a level, open site well removed from nearby obstacles. Average speed and direction are estimated by eye from dials or calculated from chart traces which provide a continuous, permanent record of the wind speed and direction.

Wind speed is expressed in km/h, although for marine and aviation purposes knots are generally used. Wind direction is the direction from which the wind is blowing. The prevailing direction is the direction most frequently observed during the day, month, or year. A gust is a brief peak wind that may occur at any time.

Local Winds

In some areas, local winds frequently blow from a different direction than that of the prevailing winds. Examples of local winds are land and sea breezes, city and country winds, winds channelled through deep fiords, and mountain or slope winds.

Land and sea (or lake) breezes, blowing from land to water at night and from water to land during the day, are well recognized in coastal areas and around large lakes and along rivers. A similar process occurs between hills and valleys, and between urban and rural areas. In urban areas during the summer, under calm or light winds, there is often a hint of a country zephyr by day and a reverse city wind by night. Hills and mountains often create a local prevailing pattern by channelling or deflecting the flow of passing winds.

A WIND BY OTHER NAMES

The tales of early peoples are resplendent with imaginative references to local winds. Some names refer to the wind's origin or destination, some describe the warmth or coldness of the air, and some tell of pleasant or trying times. Other names mimic wind sounds or suggest wind effects. Although some names have endured, others have been dropped, forgotten, or changed.

The following list, by no means complete, is a sample of local Canadian winds.

Barber: a strong wind which brings precipitation that freezes upon contact, especially with the beard and hair.

Blizzard: a cold gale-force wind laden with blinding powdery snow that reduces visibility to under 1 km. Blizzard winds usually persist for longer than six hours.

Chinook (Snow Eater): a dry warm wind that blows down the eastern slopes of the Rockies in North America.

Cold Marker: a cold north wind that creates ground snowdrifts.

Dust Devil: a rapidly rotating column of air about 30 to 100 m or more in height that picks up dust, straw, leaves, or other light material.

Flaw or Scud: a sudden gust of wind.

Nor'easter: a moderate to strong wind that blows from the northeast across the Atlantic Provinces. It is usually part of a deep, migrating low pressure system with extensive cloud and precipitation.

Plow Wind: a strong drowndraft that appears to blow out of the clouds. It is associated with squall lines and thunderstorms and has the force of a tornado.

Squamish: a violent out blowing of winds from Squamish along Howe Sound and through the channels around Bowen Island, British Columbia. It is caused by the presence of a strong, cold, wintertime high pressure system centred over the interior of the province.

Suete: a strong wind along the west coast of Cape Breton Island, Nova Scotia.

Taku: a hurricane-force wind that blows down the Taku River Valley in British Columbia.

Wreckhouse Effect: an extremely strong wind in western Newfoundland between Port-aux-Basques and Stephenville. These winds have been known to blow trains off tracks and trailer trucks off the Trans-Canada Highway.

Yoho Blow: a strong, cold wind localized in the Yoho Valley between Kicking Horse Pass and Field, British Columbia. It occurs when there is a general outbreak of cold air from the west.

LAND AND SEA BREEZE CIRCULATION

DAY

Air heated over land and rising

Air cooling and descending

Cooler air over water moving landward

NIGHT

Air cooling and descending

Cooler air over land moving seaward

Air over water rising

fog...
THE THICK OF IT

Fog is one of our more common hazards and one of our expensive inconveniences. Thick pea-soupers play havoc with air, sea, and ground transportation. Millions of dollars are lost each year when airplanes are diverted to alternative airfields. In 1973, thousands of Christmas holidayers were stranded in southern Ontario when fog blanketed the region for days. Dense fog makes navigation on waterways difficult and treacherous. A patch of river fog may have contributed to the collision on May 29, 1914, between the CP ocean liner *Empress of Ireland* and the freighter *Storstad* on the St. Lawrence River east of Québec City. Over 1000 passengers were lost. Fog also slows road traffic, and multi-vehicle pile-ups frequently occur on fog-enshrouded highways.

What is Fog?

Fog is simply a cloud that touches the ground. As a cloud, it is composed of many millions of tiny, visible water droplets. Meteorologists report thick fog when the visibility is less than 1 km, and they define a day with fog as one on which thick fog occurred for part of the day or for the entire day.

Fog forms when air is cooled to the saturation point, the temperature at which invisible water vapour begins to condense into minute liquid droplets. The requirements for fog formation are the presence of a moist air mass, a cooling process, condensation nuclei (minute particles upon which water vapour can condense), and

light winds. Cooling occurs when strong night-time radiation from the earth's surface cools the air near the ground; when humid and relatively warm air moves across a colder land or water surface; when moist air moves up and over higher terrain; and when cold rain falls through a mass of relatively dry air, thereby increasing its moisture content and lowering its temperature at the same time.

Fog Frequency

THICK AND THIN OF IT

Few areas in Canada escape being fogged in at least a few times every year. The least foggy area is in the dry B.C. interior. Penticton, for example, with an average of 1 to 4 fog days a year, receives less fog than any other station in Canada. Another relatively fog-free area is to be found in the continental middle of Canada, from Labrador in the east, across central Quebec, northern Ontario, and the Prairies to Yukon. The few fogs that do occur in this zone are

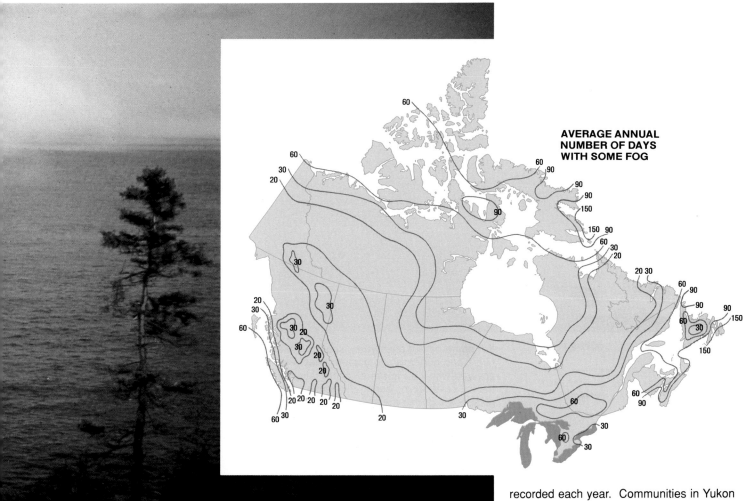

AVERAGE ANNUAL
NUMBER OF DAYS
WITH SOME FOG

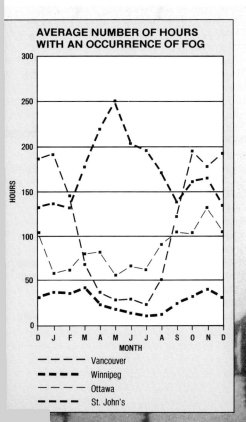

**AVERAGE NUMBER OF HOURS
WITH AN OCCURRENCE OF FOG**

- - - - Vancouver
▬ ▬ ▬ Winnipeg
– – – Ottawa
▪ ▬ ▪ St. John's

early morning events and are usually gone by noon because of evaporation and warming by the morning sun.

The foggiest area in Canada and one of the foggiest in the world is Newfoundland's Avalon Peninsula. In an average year, fog beclouds Argentia and Belle Isle for more than half the year. Argentia records about 206 days of dense fog annually, but in 1966 there were 230 days of fog. Fog is frequent along the Nova Scotia coast and the Bay of Fundy, and in the Gulf of St. Lawrence, all of which experience about 120 fog days. When the prevailing winds are southerly, sea fog drifts ashore from the cold Atlantic waters, particularly in spring and summer, and more days are counted as foggy than fog-free.

Two other areas with a high fog frequency are the Arctic and Pacific coasts. All along the Arctic coast shallow sea fog drifts ashore, especially during the warm months. The worst conditions occur in Hudson Strait and the southeastern headlands of Baffin Island, where more than 120 fog days are

recorded each year. Communities in Yukon and the western Arctic are especially prone to ice fog during periods of very cold weather with calm or light winds. Ice fog is formed at temperatures below -30°C when water vapour condenses directly into tiny ice crystals. With the added moisture pumped in from industrial sources and vehicle exhausts, dense ice fog forms more frequently and can linger for days.

Although the Pacific coast is notorious for its lengthy spells of foggy weather, the fog does not often extend far inland. Protected bays and inlets escape with only 30 days of fog a year on average. Vancouver is an important exception. There the combination of light winds, ample moisture, mountains, and local pollution sources gives rise to extensive fog. About 14% of all hours in a year have fog, as do about a quarter of the midwinter hours.

Southern Ontario and southwestern Quebec are other fog-rich areas, with some places recording 60 or more days a year. Near the Great Lakes, spring and summer is the favoured time. Inland, it is usually more common in late summer or early fall, when radiational cooling creates early morning fog problems.

ice...

THE THIN OF IT

For some part of each year, all of our lakes, rivers and northern and eastern seas are covered with ice. In the southern latitudes, the ice cover forms during the winter and disappears during the spring, but in the Arctic, waterways may never become completely ice-free.

Though "frozen water" brings pleasure to skaters and ice fishermen and provides essential winter transportation over ice bridges and roads, ice also interferes or prevents a wide variety of activities. On the Great Lakes, navigation, the regulation of lake levels, and the maintenance of hydraulic works and power strengths depends on a knowledge of both the presence and movement of ice. The area and thickness of ice may also be important in controlling the hatching and early development of certain fish species and in determining to some extent the occurrence of lake-effect snowstorms in midwinter.

Along the East Coast, ice is a major cause of damage, resulting in loss of vessels and equipment, loss of life, and major ecological disasters. In the Arctic, ice is one of the most visible phenomena in the region. It restricts ship transit and exploration but, because of its enormous load-bearing capacity and strength, supports oil drilling platforms and large military and commercial aircraft.

Inland, ice is a major factor in controlling the flow of rivers. Floods associated with river ice jams are another feature of the snowmelt — break up season. Knowledge of the duration and severity of the freeze-up and breakup seasons on inland lakes and rivers is important for making decisions as to when to move personnel and machinery across the North to lay pipelines, construct winter roads, and operate logging and mining camps.

River and Lake Ice

Most lakes in central and northern Canada freeze over completely; rivers may or may not, depending on their location, size and flow. In general, rivers freeze over later and clear earlier than lakes in the same area. Once the mean daily temperature drops below freezing, medium-size rivers become ice-fast in 4 to 5 weeks; moderate-size lakes take 2 to 3 weeks to freeze over; smaller rivers and lakes freeze in under 2 weeks. Large rivers like the Mackenzie and Upper Fraser and large lakes such as Great Bear and Great Slave require 6 weeks or longer.

Break up normally begins when the mean air temperature rises above 0°C. Ice-clearing time on rivers varies from 1 week on small rivers in southern Canada to more than 4 weeks for the Upper Mackenzie. Many Arctic rivers may never become completely ice-free because of the short melting season. For lakes, clearing times range from 2 weeks for small, shallow southern lakes to more than 8 weeks for large, deep northwestern lakes.

Great Lakes Ice

The large, deep, temperate Great Lakes have a much shorter ice season than the adjacent smaller lakes. Ice first forms in December in bays, inlets, straits, and elsewhere where shallow waters rapidly lose heat to the cold air. Maximum ice coverage occurs during late February or in March. Break up from pack ice to slush to open water is a much more rapid process. During

March the ice cover begins to clear, although ice may still be present in mid-May.

With its enormous heat storage and wind and wave action, Lake Superior is completely covered with ice only in the severest winters, and even then full coverage is likely to be of short duration.

Ice formation over Lake Huron and Georgian Bay begins in the shallower waters of the North Channel. The percentage of maximum lake surface covered with ice in a mild, normal, and severe winter is 40, 70, and 95%, respectively.

Being very shallow, Lake Erie freezes earliest and most completely and clears soonest. In the mildest winters, it is one quarter covered. Each spring, prevailing westerly winds and currents push ice into the eastern end of Erie causing ice jams that require freeing by ice-breakers.

Lake Ontario's small area and great depth give it a large heat storage capacity. Even in a severe winter, only a quarter of the surface has ice. Only twice in the past 125 years has Lake Ontario come close to being completely frozen over.

East Newfoundland Ice

In November, ice begins its drift from the Baffin coast southward to the northern Grand Banks. At the same time, new ice develops in the sheltered bays and inlets, especially those with fresh water outflow, of the Labrador coast. After a summer when Baffin Bay does not clear completely, the risk of very thick, hard, old ice floes appearing off Newfoundland the following spring is greater than in normal years. In normal years, the edge of the ice area is located near the Strait of Belle Isle by January 1. By the end of the month the mean ice edge has advanced southwestwards to near western Notre Dame Bay, and by the end of February it has progressed to about 48°N latitude. The ice reaches its maximum southern extent in March when it is at the latitude of St. John's. The pattern, extent, and concentration of ice are different during severe winters, such as 1971-72, and mild winters, such as 1957-58. Persistent winds also affect ice characteristics and extent. Offshore winds spread the ice seaward and generally lower concentrations. Onshore winds fill the bays with hummocky pack ice, which endangers fishing equipment. In early spring, while the ice recedes, ice concentrations begin to drop even though the ice cover remains substantial. By mid-May, the Strait of Belle Isle is usually navigable, and by late June, the mean ice edge has retreated to the mid-Labrador coast.

Gulf Ice

Most ice in the Gulf of St. Lawrence forms within the Gulf itself, although a smaller amount arrives from the north through the Strait of Belle Isle. The ice is constantly moving, and winds and currents eventually carry much of it into the Atlantic Ocean through Cabot Strait. From late February to mid-March, ice in the Gulf is near its maximum thickness and extent. The southwestern area is fairly shallow with weak currents. As a result the ice cover there is more stable and the floes are larger in areas where stronger currents cause congestion and pressure in the ice-pack.

Gulf ice begins to disperse in early March, starting in the estuary near the mouth of the Saguenay River. Tidal activity, increased solar warming, and higher air temperatures and precipitation rapidly erode the ice cover. In early April, the estuary is usually free of ice, and by mid-May enough ice has decayed or drifted eastward to leave the southern Gulf clear.

Icebergs

Nearly all icebergs that drift southward along the Labrador and Newfoundland coasts were calved from west Greenland glaciers. Each year between 20 000 and 40 000 icebergs plunge into the icy waters and begin their slow counter-clockwise loop around Baffin Bay. Two or three years later, only a fraction of the medium-sized bergs have managed to escape melting, calving, or grounding and reach the northern Grand Banks. Current drift and, to a lesser extent,

VISUAL SATELLITE IMAGERY OF THE GULF OF ST. LAWRENCE FEBRUARY 18, 1992

ENVIRONMENT CANADA - PHIL COTE

ICE COVER EVENTS - GREAT LAKES

Ice events during normal winter	Superior	Michigan	Huron	Erie	Ontario
Early Ice Cover	Jan 20 — 30	Jan 25 — Feb 5	Jan 25 — Feb 5	Jan 15 — 25	Jan 25 — Feb 5
Mid-Season Ice Cover	Feb 25 — Mar 5	Feb 20 — 28	Feb 25 — Mar 5	Feb 1 — 10	Feb 15 — 25
Maximum Ice Cover	Mar 25 — Apr 5	Mar 15 — 25	Mar 20 — 30	Feb 20 — 28	Mar 10 — 20
Early Decay Period	April 1 —10	Mar 20 — 30	Mar 25 — Apr 5	Feb 25 — Mar 5	Mar 15 — 25

Percentage of Maximum Lake Surface Ice Covered					
Mild Winter	40	10	40	50	8
Normal Winter	60	40	60	95	15
Severe Winter	95	80	80	100	25

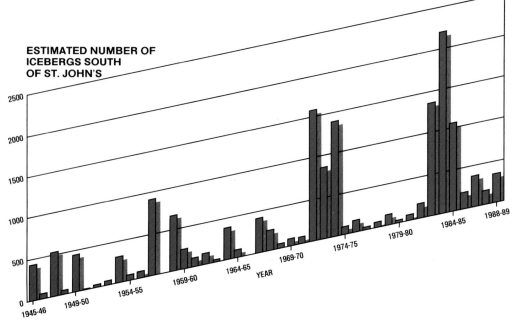

ESTIMATED NUMBER OF ICEBERGS SOUTH OF ST. JOHN'S

(Graph: x-axis labelled YEAR, with markers 1945-46, 1949-50, 1954-55, 1959-60, 1964-65, 1969-70, 1974-75, 1979-80, 1984-85, 1988-89; y-axis 0, 500, 1000, 1500, 2000, 2500)

were started in 1912 range from 0 in 1966 to 2202 in 1984.

Arctic Ocean Ice

Each winter, ice covers most of the waterways of the Arctic Archipelago, Baffin Bay, the Beaufort Sea, and Hudson Bay. In fact, many bays and inlets on the Arctic Ocean remain completely or mostly covered by sea ice for 9 to 12 months of the year. The ice cover in the Arctic Ocean is composed mainly of ice floes that are more than a year old and that range in size from a few metres to 10 km or more. Southern areas have mostly first-year or younger ice. Ice thickness varies from 5 to 30 cm for new and young ice, from 30 to 200 cm for first-year ice and from 2 to 4 m for old ice. In some cases, multi-year ice can grow to as much as 5 to 8 m thick, although growth much beyond that thickness is limited owing to the insulating properties of the ice. Pressure ridges and hummocks formed by floes that have been forced together can measure 20 to 30 m in thickness.

First-year ice predominates in the eastern Arctic from Nares Strait (between Greenland and Ellesmere Island) to Davis Strait in the south. By mid-July, Baffin Bay is sufficiently open to permit easy access to Labrador in the south and Jones Sound to the northwest. Before freeze-up begins again, little if any ice remains in Baffin Bay, having either melted or drifted out through Davis Strait. In eastern Arctic waters, tidal surges, strong northerly winds, and persistent southward flowing currents create water openings or leads.

Waters in the northernmost passages of the Queen Elizabeth Islands, the Arctic Ocean, and the western part of Parry Channel are clogged with old, multi-year ice for much of the year. A notable exception occurs in some of the deep fiords and narrow waterways between mountainous coasts, such as Eureka Sound and Greely Fiord, where ice generally clears by early August.

Along the continental coast from Amundsen Gulf to Rae Strait, clearing begins from west to east and is usually complete by late summer. The greater part of the Beaufort Sea, however, is covered with extensive multi-year floes throughout the year. These are more or less in continuous motion, the amount of movement depending on wind and current conditions. Along the mainland coast, first-year ice extends 20 to 50 km offshore. By early July it loosens and breaks up, opening a narrow channel of open water that remains through August and September. In severe ice years, the channel may be open less than a month. In a good year with mild temperatures, offshore winds, and strong tides, open water may occur for up to four months.

winds move an average of 250 icebergs into the 10-15°C waters of the Labrador Current and the Grand Banks. These majestic wintry ghosts worry mariners more than pack ice, chasing drilling platforms off site or barricading fishermen in Newfoundland's many bays and harbours. While 250 is an average, the yearly extremes since records

JOHN DEARY

ICE ISLANDS

Ice islands are enormous fragments of glacial ice calved off the Ellesmere ice shelf and set loose in the Arctic Ocean. For years, they slowly circle the North Pole in a clockwise current. The fragments may be up to about 250 km in area and have a maximum thickness of about 50 m. If the islands move into Baffin Bay, their southward drift is hastened and the fragments are soon carried to their destruction in warmer waters. Ice islands are like ships at sea but without the facility to steer. Modern northern expeditions, like the Canadian CESAR experiment in 1983, make good use of ice islands. Air strips are constructed to handle supply planes. Semi-permanent wooden buildings are erected to house a large array of monitoring and laboratory equipment, and geological and weather instruments, enabling scientists to work year-round.

IT'S THE Humidity!

HANS VANLEEUWEN

The air always contains water vapour. Even at -50°C there is still a tiny amount of moisture in the air and this can fall in the form of very small snow crystal — it can never be too cold to snow!

The wetness, dampness, or humidity of the air is one of the basic concepts in climatology, important for considering human and animal comfort, forecasting cloud and fog, determining heating and air-conditioning requirements, and assessing the condition of crops and forests.

Variations in humidity can affect the rate of evaporation. Low humidity increases evaporation because of the greater ability of the air to accept water vapour. That is why humidity levels have such an effect on our comfort in hot weather. When the air is fairly dry, the atmosphere absorbs a great deal of moisture from the skin. As a result, we feel comfortable even though the temperature may be as high as 35°C. But if the humidity is high, the moisture-rich atmosphere will extract much less water from the skin. Cooling of the skin and blood is greatly diminished, and we feel uncomfortable.

Crops are also affected by changes in air humidity. Having a spell of five consecutive rain-free, windy, low humidity days is important in drying hay and other products so as to increase their storage life. For spores and fungi that attack crops, low or high humidities can mean the difference between dormancy and activity. Similarly, knowing the atmospheric moisture content is vital in forest protection — for assessing the spread of fire and the flammability of cover and for determining the forest fire index.

Warm air can hold more water vapour than cold air. For example, air at 30°C can contain up to 30 grams of water vapour per cubic metre of air, about four to five times as much water vapour as cooler air at 5°C. Thirty grams, which is the weight of five loonies ($1 coin), may not seem like much water, but rain clouds with that much water vapour per cubic metre have the potential to produce rainfall several millimetres deep.

When the air contains the maximum possible amount of water vapour it can hold at a certain temperature, it is said to be saturated. Beyond saturation, condensation of the surplus water vapour usually starts, giving rise to clouds, mist, or fog.

Units of Humidity

Water vapour exerts a pressure which corresponds to the amount of water in the air. By calculating the vapour pressure, meteorologists can determine the humidity of the air and express it as a reading in kilopascals (kPa). Most of us, however, are more familiar with the concept of relative humidity. This is the ratio of the amount of water vapour actually in the air at a certain temperature to the maximum amount of water vapour the air could hold at the same temperature. If the air is holding only one quarter of the water vapour it could hold, then the relative humidity is a very dry 25%.

Relative humidity, however, gives very little information about the actual moisture content of the air, since it varies with the air temperature even though the amount of moisture actually in the air may not change. To avoid this problem, we can also express humidity by means of a dew-point temperature (in °C). The dew-point is found by cooling the air until it reaches the temperature at which it is saturated by the water vapour already in it. This temperature is called the dew-point because it is the temperature at which dew will form.

Humidity Across Canada

Because the air's moisture capacity varies with its temperature, the air is driest in midwinter and moistest in summer. Over the course of a day, variations in humidity are usually slight, except along the coasts where sea breezes can bring a sudden flow of moisture-laden air.

Humidity tends to be substantially higher in coastal regions and other wet areas and lower inland and in dry areas. British Columbia's interior valleys and the southern Prairies have some of the driest air in Canada throughout the year. The Arctic, though, is even drier. Nearer the Great Lakes the humidity increases. The highest values in this region occur in southwestern Ontario, where humid tropical air is often present during the warm months. In winter, the West Coast has comparatively high humidities. In the Maritimes, winds associated with relatively dry, continental air masses and nearby cold ocean currents tend to keep humidities low. Local humid spots can occur over hills when low cloud reaches the ground or in valley bottoms when mist or fog forms on cold nights.

JANUARY AVERAGE VAPOUR PRESSURE (kPa) AT 1 P.M.

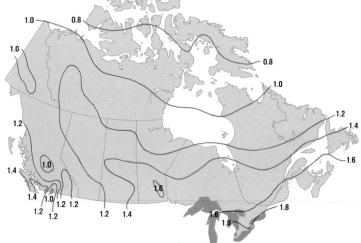

JULY AVERAGE VAPOUR PRESSURE (kPa) AT 1 P.M.

TYPICAL HUMIDITY VALUES RELATIVE HUMIDITY IN % VAPOUR PRESSURE IN KPA							
	Winter				Summer		
	a.m.		p.m.		a.m.		p.m.
	%	kPa	%	kPa	%	kPa	% kPa
West Coast	85	0.63	80	0.60	90	1.35	75 1.40
Dry Inland	80	0.15	75	0.20	75	1.35	75 1.35
Lower Great Lakes	85	0.30	80	0.30	90	1.80	60 1.80
East Coast	85	0.40	75	0.45	90	1.60	70 1.70
Arctic Island	70	0.05	70	0.05	85	0.70	80 0.70

EVAPORATION

Almost three quarters of the rain and snow that falls upon Canada is returned to the atmosphere through evaporation. Water recycles between the air and ground several times whiile an air mass makes its way across Canada.

Evaporation is the reverse of precipitation in the water cycle, and over the entire planet, the water involved in the two processes must balance. Evaporation occurs directly from a free water surface or by transpiration from vegetation, a process by which water is drawn from the soil by the root system and is passed into the atmosphere in the form of water vapour through tiny openings in the plant's tissues.

Precipitation received minus moisture lost through evaporation and transpiration (together known as evapotranspiration) determines the water available for several uses. Evaporation removes immense quantities of precious water from the soil and from storage facilities such as reservoirs. Lake levels rise and fall depending on the net supply of water after precipitation and evaporation. Evaporative losses affect the quantity of water available for domestic and industrial use and for irrigation. What is more, evaporation has almost as much influence upon crop production as rainfall does.

With every molecule of water evaporated into the air, there is an associated transfer of heat. This energy fuels our storm systems.It is for this reason that a knowledge of the distribution of evaporation across the globe is basic to the understanding of the general circulation of the atmosphere. In order for evaporation and transpiration to occur, there must be not only an adequate water supply but also a sufficient energy supply to change the water from a liquid to a gas. This energy is derived from the sun. The energy that is used in evapotranspiration is, in fact, stored by the atmosphere and is released when the water vapour changes back from a gas into a liquid or solid form as rain or snow.

Generally, for vegetation and shallow water bodies, the rate of evapotranspiration will closely follow the daily and seasonal cycles of solar radiation, with the maximum rate occurring in the summer. However, for large water bodies such as the Great Lakes, much of the solar energy in the spring and summer is used for heating their large water mass rather than for evaporation. As a result, maximum evaporation from these large water bodies occurs later, in the fall, with much of the required energy coming from the heat stored in their water mass.

Lake Evaporation

Some simple methods have been developed to estimate the amount of water lost by water bodies through evaporation. One such method is the use of evaporation pans. The evaporative loss from these pans can be adjusted to provide an estimate of "lake evaporation," that is, evaporation from small natural water bodies, such as reservoirs and ponds, in which heat stored by their water mass is negligible. Evaporation and evapotranspiration losses are most commonly measured in millimetres of depth lost over a specified area.This enables easy comparison with precipitation measurements, which also use these same units.

Lake evaporation is greatest in the dry Prairies, where it can be as much as 1000 mm a year. It is least, however, in the even drier lands of the extreme north, where it averages around 100 mm. Evaporation losses here are minimized because lakes are snow- and ice-covered for nine or more months of the year.

Most areas have an evaporation season (coinciding with the frost-free season) from May through October. The Pacific coast has the longest season, covering 12 months in some places. Annual losses here run from 400 to 700 mm. At high elevations in the Cordillera, the season is short and lake evaporation is low because of the colder temperatures and longer ice season. July is the month when maximum evaporation occurs in the south, and August is the peak month in the Arctic islands. On the Prairies evaporation losses are also high in May. In fact, they are higher than in June, which is traditionally the cloudiest and wettest warm season month.

Lake evaporation across eastern Canada ranges from about 400 mm in the north to 800 mm in southwestern Ontario. In the east there is a regular seasonal cycle with a fairly well-marked maximum in July.

AVERAGE ANNUAL LAKE EVAPORATION (mm)

200 400 800

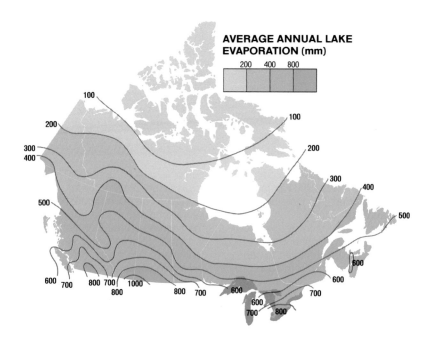

Too Much Water

Few natural hazards have been so well documented throughout history as floods. Every year, since the Great Deluge in biblical times, flooding somewhere has caused enormous destruction and human suffering, even in areas where high water is a normal occurrence. Most floods are caused by too much rain in too short a time.

But in Canada, there are several other causes of floods besides storm rainfall. These include snowmelt, ice jams, wind-driven water, and high tides.

Hurricanes and large, slow-moving low pressure systems may produce long periods of continuous heavy rain sufficient to cause flooding. However, flooding from such

storms is rare in Canada. The most severe occurred on October 14-15, 1954, when Hurricane Hazel brought 214 mm of rain to the Toronto region in a period of 72 hours. The flooding on the Don and Humber rivers and the Etobicoke Creek resulted in 80 deaths and $25 million (100 million in 1988 dollars) in public and private property damage.

Along the East Coast, low pressure systems set in motion enormous storm surges or great domes of water often 80 or more km wide. The water surge adds to the normal high tide, and battering waves produce a storm tide that poses a serious threat to coastal property. Coastal flooding may also be caused when a *tsunami*, a huge wave generated by underwater earthquakes, is driven ashore.

Flash flooding is a problem in Canada, although because of differences in population density and topography the consequences are usually not nearly as severe as in the United States. Flash floods can occur when an intense thunderstorm drops a deluge on a fairly small area in a very short space of time. Intense, short-duration rainfalls cannot be absorbed by the ground or handled by the system of storm sewers. Urban areas, where there are more impervious surfaces, are especially vulnerable. Sewers burst and basements fill and the rush of water causes widespread damage and even loss of life. On July 10-11, 1883, an intense downpour of 77 mm raised a "wall of water" that descended on London, Ontario, carrying away buildings and drowning 18 people. In 1981 a thunderstorm dropped 56 mm of rain in less than 3 hours on Edmonton. Flooding was widespread,

FLASH FLOOD

SNOWMELT

PETER ELMS

ICE JAM

JIM OLEKSUK

WIND-DRIVEN

ENVIRONMENT CANADA

STORM SURGE

CHRIS STEWART

washouts and mudslides occurred, and four workers were drowned. The most spectacular rainfall event in Canada occurred at Buffalo Gap, Saskatchewan, on May 30, 1961, when 250 mm of rain fell in less than one hour. An even greater amount fell in 18 hours over parts of Essex County in southwestern Ontario on July 19-20, 1989, inflicting $35 million damage to one of Canada's prime agricultural areas.

For much of Canada, spring is the peak flood season because of snowmelt and ice jams. The combination of heavy winter snowfall, late spring thaw, and heavy rains during the melt period may produce extensive damage in flood-prone areas from Vancouver to St. John's. Perhaps the most dramatic example of snowmelt flooding occurred in 1948 along the Fraser River. More than 10% of the Fraser Valley was under water. Ten persons were drowned and thousands left homeless, with property damage totalling $30 million. In 1950, about one fifth of Winnipeg was under water and 100 000 people were evacuated during the great Red River flood. While damage costs were high, there was no direct loss of life.

On some watercourses the rush of runoff to the rivers during an early thaw can occur while the rivers are still frozen over. As soon as the ice breaks up and begins to move, the danger from ice dams appears. Ice floes jam at constrictions in the river, forcing water levels to rise rapidly upstream by as much as a metre in minutes. Flooding due to ice is a problem along the larger northward-flowing rivers like the Red, the Chaudière, and the Mackenzie.

The levels of lakes and other inland water bodies rise rapidly from excessive precipitation and river runoff. During stormy, windy weather, considerable shoreline damage may result. Under certain conditions, strong winds can tilt the lake surface by piling up water on one shore and causing a drop on the opposite shore. Fluctuations of 3 to 5 m may last for as long as 12 hours before the water sloshes back and forth to a steady level.

GREAT CANADIAN FLOODS

1798	Floods at Montréal and Trois-Rivières, Quebec, were described in contemporary reports as the "worst in living memory."
1826	The Red River in Manitoba rose 2 m higher than during the famous flood of 1950.
1865	The St. Lawrence River rose 3 to 4 m at Sorel and Trois-Rivières, Quebec. 45 drowned.
1883	A wall of water along the Thames River drowned 18 in London, Ontario.
1928	The Rideau, Chaudière, and Quyon rivers in Quebec overflowed their banks. Several drowned.
1937	The Thames River at London, Ontario, flooded, leaving 4000 homeless.
1948	Floods in the Fraser Valley in British Columbia left 9000 homeless. This flood has been called the "worst in B.C. history."
1950	The Red River in Manitoba rose 10 m above normal. 100 000 were evacuated.
1954	The Etobicoke Creek in Ontario flooded during Hurricane Hazel. 80 died.
1973	The Saint John River in New Brunswick flooded, causing $12 million damage.
1974	The Grand River flooded Cambridge, Ontario, causing $7 million damage. Extensive flooding across the Prairies.
1979	The Red River in Manitoba crested higher than in 1950. The Yukon River flooded 80% of Dawson, Y.T.
1980	Boxing Day floods along the Squamish River in British Columbia caused $13 million damage.
1986	The rain-swollen North Saskatchewan River rose 7.6 m above normal levels and forced 900 Edmonton residents to flee their homes.

Too Little Water

MRS. R. WEBBER

Drought is a hydrometeorological riddle. We can measure its effects but we can't always discern its presence. Farmers define it differently than suburbanites and water engineers. It creeps slowly upon us but ends with incredible abruptness—and often in floods.

There are droughts and there are dry spells, and the difference between the two can often be one of personal opinion. Droughts can be permanent or invisible, and there are various degrees of drought severity and duration. There are no universally accepted definitions of drought. Any extended dry weather that is worse than expected and that leads to measurable losses can correctly be called a drought.

Drought is a chronic concern in Canada, frequently threatening water supplies, wetlands, and forage and cereal production. It creates conditions favourable to the spread of certain insect pests and plant and animal diseases. Drought also creates ripe conditions for forest fires.

Few regions in the world are immune from the hazard of drought. Droughts in the African Sahel and northeastern Brazil are almost expected. Even moist areas of the northeastern United States and of Great Britain feel the scourge—but not in the same devastating way as countries in the

Third World.

In Canada the effects of drought are felt most in the Prairies, especially in the arid portion of southern Saskatchewan and Alberta known as the Palliser's Triangle, where annual precipitation is under 350 mm. In most years the Prairie climate is favourable for grain cultivation, but from time to time, and on occasion several years in a row, the regional climate reverts to an arid type.

Drought is usually of less concern in the dry interior valleys of British Columbia, where irrigation is practised. A notable exception occurred in 1987 when precipitation averaged only 60% of normal yearly accumulations. Droughts also occur in Eastern Canada, although the hazard is usually short-lived, relatively local in extent, and less frequent and less severe than on the Prairies.

Famous Canadian Droughts

Recurrent drought is inevitable across the Prairies. During the past two centuries at least forty serious droughts have occurred in Western Canada.

Unmistakably the most calamitous was the drought of the 1930s when a quarter of a million people abandoned the West. During the dry years between 1933 and

DROUGHT YEARS

1805	scorched potato crop in the Red River area
1816-19	almost continuous drought and hordes of grasshoppers on the Prairies
1846	complete crop failure in the Red River area
1862-4	low river levels on the Red River; seed grain imported from USA
1868	Prairie crop fails; plague of grasshoppers
1890s	9 years of drought force farm abandonment
1933	grasshopper plague and drought result in the smallest wheat crop in Saskatchewan since 1920
1936	severe heat stress in Ontario reduces crop yields by 25%
1936-8	recurrence of drought on Prairies a national emergency
1961	the worst drought year this century for Prairie wheat
1963	severe Ontario drought drastically cut soybean and corn production
1967	extensive drought from Peace River to southern Manitoba
1973	record warm summer and local drought hurt potato and apple production in Ontario
1977	severe drought in southern Alberta and western Saskatchewan
1978	extensive central Ontario drought
1979-80	two poor wheat and pasture years on the Prairies; estimated loss to national economy $2 1/2 billion
1983	southern Ontario drought described as worst this century
1984	drought in southwestern Nova Scotia dried up many streams and wells
1985	southeastern Prairies received half normal amounts of rain; insect infestations; forest fires in British Columbia cost $300 million (firefighting costs and timber losses)

1937, the Prairies collected only about 60% of the normal rainfall. The years 1936 and 1937 were especially difficult. Thousands of head of livestock were lost to starvation and suffocation, and crops were withered and stunted.

Numerous other very serious but relatively short-term droughts have plagued the West. The single driest year was 1961. In the heart of the drought-stricken area only 45% of normal precipitation was recorded at some locations. At Regina every month but May was drier than normal, and for the 12-month crop year the precipitation total was the lowest ever.

The duration, severity and size of the area effectively made this drought the worst on record. Losses in wheat production alone were $668 million, 30% more than in the previous worst year, 1936. Water was rationed, livestock herds were culled, and hydroelectric production was curtailed. The drought had a rippling effect on the national economy.

The 1980s have rivalled the 1930s in terms of the intensity and duration of the dry spell and the extent of the extremely dry area. The year 1988 was especially disastrous for Prairie grain farmers and ranchers. For many localities, it was the hottest summer on record. Growing-season rainfalls across the south averaged between 50 and 80% of normal rainfalls, although the dry period really began in September 1987 and continued through the largely snow-free winter and record dry spring. When the rains finally came, they arrived at the worst possible time, in late August and September when frustrated farmers were trying to harvest the drought-ravaged, insect-infested crops.

The effects of the 1988 drought were disastrous for nearly all segments of the Western economy. Agriculture was particularly hard hit. Grain production was down by an average of 31% from 1987's, and export losses were estimated at 4 billion dollars. Ranchers were forced to thin their herds or move their cattle long distances to find adequate grazing land. As a result of their economic difficulties, about 10% of farmers and farm workers left agriculture in 1988.

Outside the agricultural sector, hydroelectric generating capacity and power exports were affected, communities faced water shortages, and wildlife was decimated. Eventually, the effects were felt across the country, as lower agricultural yields led to higher food and beverage prices for consumers.

THE STORMS OF SUMMER

ENVIRONMENT CANADA

AVERAGE ANNUAL NUMBER OF DAYS WITH THUNDERSTORMS

5 10 20

Thunderstorms

The thunderstorm is one of the most familiar and characteristic features of summer weather across most of southern Canada. The drama unfolds, typically on a warm, humid day, while fair weather cumulus clouds build into towering cumulonimbus with white billowing tops and darkened bases. While the storm approaches, the sky darkens, the air stills, and distant thunder rumbles. Within minutes, a few drops signal the beginning of nature's most spectacular and deadly sound and light show. Torrential rain, violent winds, and damaging hail may accompany the thunderstorm, and if it is severe enough tornadoes may occur. With the same suddenness, the storm ends. The skies brighten, and only fading peels of thunder roll in the distance.

The highest average number of thunderstorm days (a day on which one or more peals of thunder are heard, regardless of the actual number of individual thunderstorms) is 34 a year and occurs in extreme southwestern Ontario, away from the Great Lakes and their stabilizing effect. A secondary peak of 25 or more days occurs between Calgary and Edmonton, in southwestern Quebec, and in southeastern Manitoba. The frequency drops rapidly northwards. In the boreal forest there are fewer than 10 thunderstorm days a year, and the Arctic Islands and the Pacific coast are largely free of thunderstorms. Elsewhere the thunderstorm count is 5 days a year over Newfoundland and Yukon, 10 to 15 over the Maritimes, 15 to 20 over central Quebec and northern Ontario, and 15 to 25 across the Prairies.

More thunderstorms occur in July than in any other month. During July, residents of Windsor and London, Ontario, can expect to have a thunderstorm every four days. April to October is the normal season for thunderstorms in central and eastern Canada, although winter thunderstorms occasionally occur in these regions. In the West, the season runs from June to August, and winter thunderstorms are extremely rare. There is a daily pattern to thunderstorm occurrence, with eastern storms peaking in the mid-afternoon and western ones commonly occurring two to four hours later. In both areas the quietest time is between 9 a.m. and noon.

Thunderstorm rain represents an important portion of the summer precipitation total. Only half the thunderstorms are rain-bearing, but extreme rainfalls almost always come from thunderstorms.

Hailers

Hail typically occurs in heavy but localized showers associated with mature thunderstorms. The main cause of hail is atmospheric instability, which often produces updraughts strong enough to carry the weight of hailstones as they grow. The greatest air instability is usually found away from large water bodies and near mountains or where intense local convection results from daytime heating.

Hail is common in Canada, but relatively rare at any single location—most southern localities report only 1 or 2 hail days each year. Small hail, the inoffensive kind, falls frequently in May and June over the Atlantic Provinces and eastern Quebec and is very common along the Pacific coast in spring. The favoured "true" hail area is the continental interior. Central Alberta's "hailstorm alley," to the lee of the Rockies, and southernmost Saskatchewan east of the Cypress Hills get hit on 4 to 6 days each year. Other parts of the southern Prairies experience 3 hail days a year, as does the central dry interior of British Columbia and southwestern Ontario.

May to July is the period of maximum hail occurrence, and nearly three quarters of all hail falls occur between noon and the dinner hours. The average hailfall usually lasts from 6 to 10 minutes.

ONTARIO DEPARTMENT OF AGRICULTURE AND FOOD

ENVIRONMENT CANADA

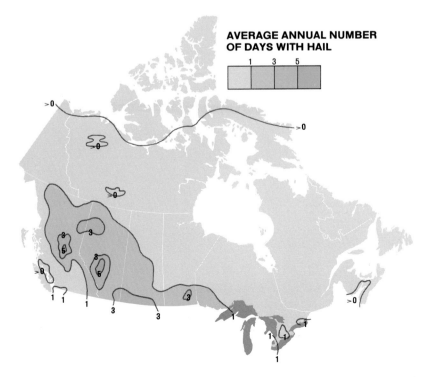

AVERAGE ANNUAL NUMBER OF DAYS WITH HAIL

1   3   5

FAMOUS CANADIAN HAILSTORMS

Calgary, July 28, 1981. A 15-minute hailstorm caused in excess of $100 million damage—a record for insurance claims in Canada due to hail fall.

Montréal, May 29, 1986. Hailstones up to 8 cm in diameter pounded the south shore. Damage estimates exceeded $65 million. One year later, another multimillion dollar hail disaster occurred in Montréal.

Windsor - Leamington, Ontario, May 30, 1985. Golf-ball size hail smashed greenhouses and flattened crops. Damage estimates were $30 - 40 million.

Western Prairies, July 23, 1971. Two days of severe weather did $20 million damage. The hail swath extended over 500 km.

Edmonton, August 4, 1969. Some of the largest hailstones ever observed in the area inflicted $17 million damage to the city and environs.

Cedoux, Sask., August 27, 1973. Damage estimates put at $10 million. One hailstone weighed 290 g and measured 114 mm across, the largest documented hailstone in Canada.

Okanagan Valley, July 29, 1946. Near Penticton a hailstorm inflicted $2 million damage to the apple and pear crop in 15 minutes.

Edmonton, July 10, 1901. Almost every tin roof was ruined and thousands of lights were broken. Hailstones measured 8 cm and weighed 140 g.

Lambeth, Ontario, August 19, 1968. A severe hailstorm left ice up to 17.5 cm deep on the streets and caused extensive damage to crops and property.

Montréal, June 5, 1979. A violent thunderstorm with hail and heavy rains caused considerable damage, especially to the Botanical Gardens.

Central Alberta, July 14, 1953. Golf-ball size hail covered 1800 square kilometres. Thousands of birds were crushed.

TORNADOES

Tornadoes or twisters are violently rotating air columns that are usually visible as a funnel cloud hanging from dark thunderstorm clouds. In area a tornado is one of the smallest of all storms, but in violence it can be the most severe. The whirling cloud moves erratically across the ground at speeds between 40 and 65 km/h. In extreme cases, winds can reach 400 km/h and pressures inside the funnel can drop to less than 90% of normal atmospheric pressure. Such awesome forces cause enormous destruction, as do the hail, rain, and lightning that usually accompany the tornado. Striking suddenly, tornadoes can demolish buildings, lift heavy vehicles, derail trains, uproot trees, and twist steel. Most tornadoes by far occur in the Great Plains region of the United States. Each year the United States experiences an average of 700 tornadoes.

Text adapted from M. Newark, Tornadoes In Canada. 1950-1979 (Toronto: Environment Canada (AES), 1983).

About 70 or 80 tornadoes a year hit Canada's populated areas, while an unknown number occur in the unpopulated forest regions. Fortunately, throughout North America more than 90% of all tornadoes that occur are too weak to cause serious damage.

Through the efforts of several thousand severe weather watchers across Canada, who report tornadoes and other severe weather sightings to the nearest weather station, and Michael Newark and his colleagues at Environment Canada, who have researched and compiled statistics on tornadoes, we now know considerably more about the climatology of tornadoes in Canada.

Where They Occur

Tornadoes have been reported in every province. They are rare west of the Alberta foothills and in the north, but the risk is high in extreme southwestern Ontario (away from the immediate vicinity of the Great Lakes), near the international border in southern Saskatchewan and southwestern Manitoba, and in northwestern Ontario. Other areas of enhanced tornado activity are northeastern Ontario, western Quebec and south-central Alberta. Low risk areas include most of Saskatchewan, north and east of the Upper Great Lakes, the St. Lawrence Valley in Quebec, and the Atlantic Region.

One third of all tornadoes reported in Canada have occurred in Ontario, where a maximum of 2 tornadoes per 10 000 km^2 can be expected annually. There is a risk of some damage to a major southern Ontario city about once every 15 years, and severe damage once in about 25 years.

When They Occur

In 1954 a tornado occurred in Nova Scotia in January. Although an occasional winter tornado does occur, records show they are extremely rare. Late June and early July are the preferred months, and from 3 to 7 p.m. is the preferred time of day. For very severe tornadoes, with wind speeds above 180 km/h, June is the peak month. On the average in Canada, the tornado season ranges from April to October — about 160 days.

In the summer, an average of 1 tornado every 5 days is reported in Canada, compared to 5 tornadoes every day in the United States.

AVERAGE ANNUAL FREQUENCY OF TORNADOES PER 10,000 SQUARE KILOMETRES

0.4   0.8   1.2

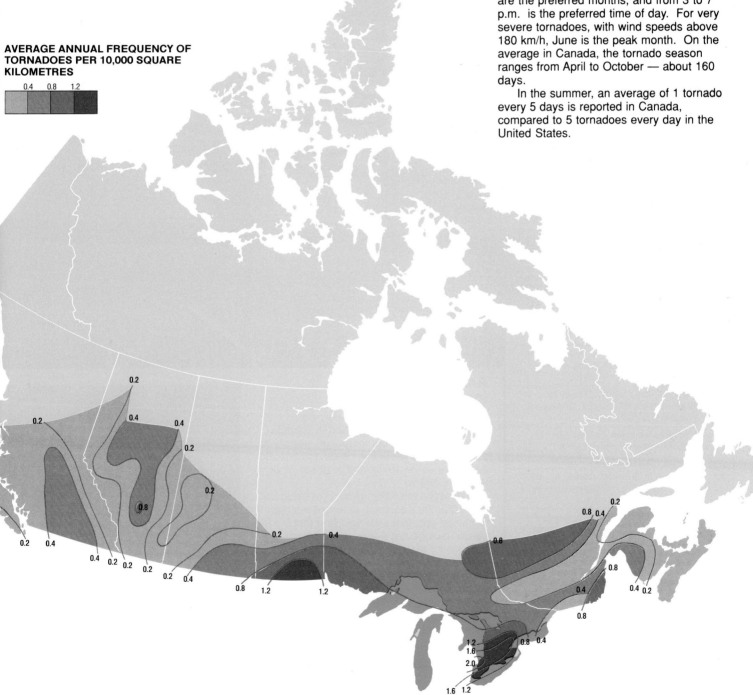

CANADA'S TEN MOST DISASTROUS TORNADOES*

1. **Regina, June 30, 1912.** The Regina "Cyclone" left 28 dead and hundreds injured. $4 million damage.

2.. **Edmonton, July 31, 1987.** 27 dead, 253 injured, hundreds left homeless. Damage was estimated at over $250 million.

3. **Windsor to Tecumseh, Ontario, June 17, 1946.** 17 dead and hundreds injured. Damage conservatively estimated at $1.5 million.

4. **St. Zotique to Valleyfield, Quebec, August 16, 1888.** 9 dead, 14 injured, extensive property damage.

5. **Windsor, Ontario, April 3, 1974.** 9 dead and 30 injured. Damage conservatively estimated at $1.5 million.

6. **Barrie and central Ontario, May 31, 1985.** 8 dead, 155 injured, $100 million damage.

7. **Buctouche, New Brunswick, August 6, 1879.** 7 dead, 10 injured, 25 families homeless, $100,000 damage.

8. **Sudbury, Ontario, August 20, 1970.** 6 dead, 200 injured, and $5 million damage.

9. **Montréal (Ste-Rose), June 14, 1982.** 6 dead, 26 injured, hundreds of homes and barns flattened.

10. **Portage la Prairie, Manitoba, June 22, 1922.** 5 dead, scores injured, extensive damage.

*Updated from Chinook, Fall 1978. Ranked from worst to least severe, based on the number of deaths.

Are Tornado Numbers Increasing?

During the 1950s, approximately 20 tornadoes were being reported in Canada annually. This number increased steadily to the 30 to 50 range in the 1960s, to about 50 each year in the 1970s and to about 70 or 80 in the 1980s.

Despite the rising trend, climatologists claim that the frequency of tornadoes has remained constant over time and that people factors, not meteorological ones, account for the increasing numbers of reported tornadoes. This is simply because there are now more people to see and report them, there are much greater densities of industry and population, and there is a greater public awareness of severe weather.

MICHAEL NEWARK

SCOTT NORQUAY

HURRICANES

Hurricanes are enormous spiralling storms containing raging winds and great decks of storm clouds that produce the heaviest rains on earth. They begin out over the warm tropical ocean as a relatively small low pressure system. Once their winds reach speeds of 120 km/h or more they attain hurricane status.

The storms are powered by the heat released into the surrounding air by condensing water vapour. The warm air rises and draws in more moisture-laden air at the surface, which rises faster and farther and releases even more heat. In time, this energy machine develops into a full-fledged hurricane with violent inward-spiralling winds, producing towering cumulus clouds and torrential rains.

The main system of cloud, rain, and wind extends about 30 km out from the centre of the eye. Farther out, stretching over tens of thousands of square kilometres, conditions are less extreme, but winds are still violent (above 120 km/h) and rains constant and heavy. The young hurricane moves very slowly in the tropics, usually advancing at a speed of 15 to 25 km/h. While it moves northward from the

ENVIRONMENT CANADA

equator its speed of movement tends to increase. But once away from the tropical zone and into cool northern latitudes, the storm tends to die out, either because it is starved for heat energy or because it is dragged apart by ground friction.

Inside the eye of a hurricane, the wind is gentle, skies are clear or dotted with puffy clouds, and the air is warm and humid. It is also the low pressure heart of the storm, with air pressure readings generally less than 96 kPa. On average, the eye is about 30 km across. Beyond the eye, through the spiralling wall of cloud, winds roar at more than 100 km/h, with maximum wind speeds exceeding 400 km/h in very extreme cases, and deluges fall continuously. Many persons have been lured out of shelter as the calm eye passes over, only to be killed or injured by violent gales, this time from the opposite side of the system.

Often the most destructive element of a hurricane is the storm tide or ocean swell. Ahead of the storm the ocean builds into a huge undulating surge that comes crashing ashore, flooding coastal areas with sea-levels as much as 5 m or more higher than normal high tides.

Hurricanes In Canada

On the average 10 tropical cyclones will develop in the North Atlantic Ocean in an area from the Caribbean Sea and Gulf of Mexico to Newfoundland each year. Only 6 of these storms will be hurricanes, and only 1 or 2 will enter Canadian waters. Although tropical storms and hurricanes have weakened considerably by the time they reach southeastern Canada, they can cause heavy rains and strong winds, usually during the period from late August to the end of October.

Hurricanes are called typhoons when they occur in the Pacific. Very seldom do typhoons stray into Canadian waters. The eastern Pacific cyclone track is located off the coast of Mexico, but the waters of the North Pacific are just too cold to sustain hurricane intensities. Typhoon Freda, or at least the remnants of Freda, struck the Pacific northeast on October 12, 1962. At Victoria winds reached sustained speeds of 74 km/h with gusts to 145 km/h. Damage estimates exceeded $10 million and there were seven storm-related deaths.

The earliest reported hurricane to strike Canada occurred on November 19, 1813. Only about one fifth of the tropical cyclones reported in Canada in the past fifty years have reached hurricane status. After travelling 1500 km, the majority of hurricanes are much less potent by the time they enter colder Canadian waters. The most affected region is the Atlantic Provinces. A small number also occur in Labrador, Quebec, and Ontario east of Lake Superior.

Hurricane Hazel is the most remembered storm in Canadian history, striking southern Ontario on October 14-15, 1954, and leaving behind 80 dead, $100 million damage (1988 dollars), up to 182 mm of rain, and wind gusts of 125 km/h. Hurricane Beth on August 16, 1971 produced more rain than Hazel, dumping 296 mm of rain on Nova Scotia. An unnamed hurricane on September 25, 1941 struck southwestern Ontario, where London reported 130 km/h winds, stronger than those that accompanied Hazel. The deadliest hurricanes ever to strike Canada were the unnamed cyclone that hit the Maritimes on August 25, 1873, when 1200 vessels were lost, and the "Independence Hurricane" that struck Newfoundland on September 9, 1775, drowning several thousand British seamen.

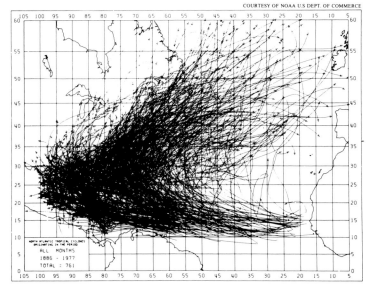

COURTESY OF NOAA U.S DEPT. OF COMMERCE

TRACKS OF 761 ATLANTIC TROPICAL CYCLONES OVER A 90 YEAR PERIOD (1886-1977)

JEAN ELLIOTT

SOME FACTS ABOUT HURRICANES

Maximum wind speed often exceeds 250 km/h.

Although the winds in a hurricane reach speeds of hundreds of kilometres per hour, the storm centre itself moves relatively slowly. The fastest moving storm on record in North America was the 1938 storm in New England, 90 km/h.

In North America the hurricane season lasts from July to November; the majority of storms occur in September.

On average, six Atlantic hurricanes occur per year. In 1985 seven tropical cyclones were named hurricanes. In 1893 four hurricanes existed at one time in the North Atlantic.

The most deadly hurricane in history was in the Bay of Bengal in 1737 when 300,000 persons were drowned.

The diameter of the eye varies from 6 to 40 km.

Rainfall from a mature hurricane may range from 80 to 150 mm.

About 39 per cent of all United States hurricanes hit Florida.

WINTER BLASTS

ENVIRONMENT CANADA

Blizzards

No region of Canada is secure from the fury of winter blasts and blizzards. Intense winter storms are frequently accompanied by numbing cold, ice or glaze, heavy snow, blustery nor'easters, or a combination of these. Nearly everyone remembers significant winter storms—days of heavy snow, slippery walks and roads, transportation delays and cancellations, extra costs, and, sometimes, personal tragedy. Winter brings them all. For the residents of New-foundland, the blizzard of 1959 was one of the worst on record. The storm, which struck on February 16, took six lives, left 70 000 Newfoundlanders without power, crippled telephone service, and blocked highways and roads with 5-m drifts.

Winter storms can kill. Deaths occur from automobile and other accidents, falls, overexertion and/or exposure, house fires,

EARL ZILKIE

carbon monoxide poisoning, and electrocu-tion from fallen power lines.

A blizzard is the most violent winter storm, combining strong winds with cold temperatures and blowing or drifting snow which reduces visibility to zero. Officially, to be classified as a blizzard, the following con-ditions must be fulfilled:

- a temperature of -12°C or less
- wind speed of 40 km/h or greater
- visibility of less than 1 km
- duration of these conditions for at least 3 to 6 hours.

Blizzards occur nearly every year in eastern Canada, but it is the Prairie "norther" and Arctic whiteout that are legendary. Tales are told of pioneer farmers who perished midway between house and barn or of Arctic explorers who died after straying only a metre or two from their average year. Newfoundland has 30 days, and southern portions of Manitoba and Saskatchewan have 20 to 30 days in an average year. The high frequency over the eastern Canadian tundra proves that heavy snowfall is not a prerequisite for blowing snow conditions. A moderate snowcover, strong winds, and a lengthy, unobstructed wind path are more contributory.

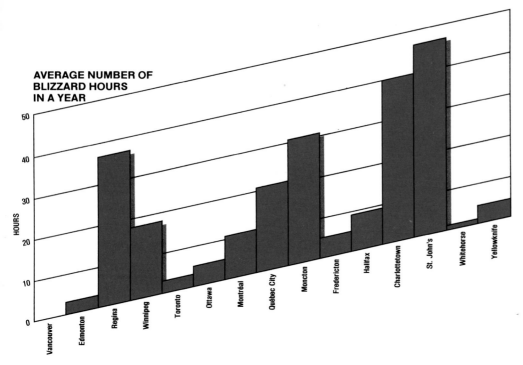

AVERAGE NUMBER OF BLIZZARD HOURS IN A YEAR

HOURS

50
40
30
20
10
0

Vancouver, Edmonton, Regina, Winnipeg, Toronto, Ottawa, Montréal, Québec City, Moncton, Fredericton, Halifax, Charlottetown, St. John's, Whitehorse, Yellowknife

campsites. Even today, not a winter goes by without tragic news being heard about travellers succumbing to a blizzard.

Blowing Snow

Blowing and drifting snow create tension and difficulties for travellers disoriented momentarily by the loss of visibility. A blowing snow day is one on which moving snow (i.e., snow lifted from the ground by the wind) reduces visibility to 10 km or less at least once during the day. Such days are common in the eastern Arctic and over open, windy areas of southern Canada east of the

Rockies. The valleys of British Columbia, southern Yukon, and the Boreal forest of the Prairies have fewer than 10 days of blowing snow a year on average. Substantially more days, up to 20, are found over southern Ontario and southern Quebec. The worst area is in the Arctic islands and over the central mainland of the Northwest Territories, with more than 90 days a year. The area around Hudson Bay has more than 60 blowing snow days. North of the tree-line from the Beaufort Sea southeast to James Bay and central Quebec, blowing snow days number 30 or more in an

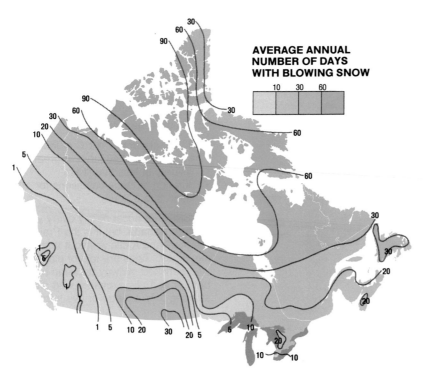

AVERAGE ANNUAL NUMBER OF DAYS WITH BLOWING SNOW

10   30   60

ENVIRONMENT CANADA

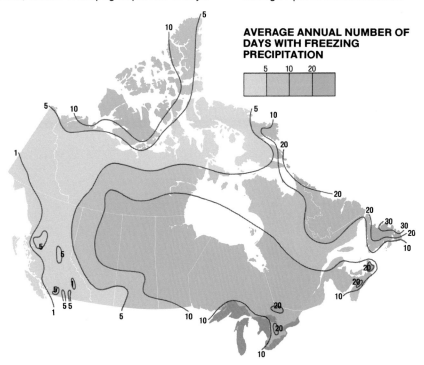

PETER ELMS

Glaze Storms

Freezing precipitation is a major hazard in many parts of Canada. Heavy icing, especially when accompanied by strong winds, devastates trees and brings down wires, disrupting communications and power supplies. Sidewalks and roadways become extremely hazardous, and ice loads on roofs may become unbearable. A severe ice storm may inflict social and economic chaos, especially in urban areas. The effects may be felt for days or even weeks afterwards. Montréal suffered one of its most damaging ice storms on February 25 — 26, 1961. Hydro wires, coated with 5 cm of ice, snapped in the strong winds, leaving some areas of the city without power for a week.

Freezing rain or drizzle begins as liquid water droplets that fall from warmer-than-freezing upper air into a shallow layer of colder air near the ground. When the surface temperatures of objects and of the ground are below freezing, the droplets freeze upon impact, glazing everything with a silver coat of ice.

Some of the most destructive ice storms have occurred in Newfoundland, where they are known as "silver thaws." Such a storm struck St. John's in April 1984, when freezing precipitation began on the evening of April 11 and continued intermittently until the 14. Over 200 000 persons in the Avalon Peninsula were without electricity for days, and a shortage of kerosene heaters occurred as people looked for other ways of heating their homes. Cylinders of ice as large as 15 cm in diameter formed on overhead wires.

Ontario has also had its share of ice storms. The most memorable one produced a 3-day period of freezing rain in January 1968. As well, southern Manitoba has been hit with several ice storms in recent years. A particularly destructive storm, on March 6, 1983, closed Winnipeg airport for 2 days and brought down several TV towers.

As might be expected, British Columbia has the least freezing rain and Newfoundland the most. Gander has 51 days a year on average and St. John's 36. Other cities, with their average number of freezing rain days, are: Vancouver 1 day, Edmonton 8 days, Winnipeg 12, Toronto 10, Montréal 13, and Halifax 19.

The riskiest months for freezing rain are November and December in the West; November, December, and January in central Canada; December through March in southern regions of Ontario and Quebec and in the Maritimes; and December through April in Newfoundland.

AVERAGE ANNUAL NUMBER OF DAYS WITH FREEZING PRECIPITATION

5   10   20

Climate Regions

In a country as vast as Canada, extending across half a continent, it is more appropriate to speak in the plural about the *climates* of Canada. Our climate is a mosaic of most of the climates of the Western Hemisphere. Canada has the snow and cold of Siberia; the storms of the United States; the summer heat and humidity of the Caribbean; the aridity of the American desert southwest; the moistness of Ireland; the winds and fog of Great Britain; and the temperateness of the Mediterranearn countries. Of the world's major climates only the desert and equatorial rainy types are absent.

There are innumerable accounts of the climate of Canada. Writers have used the following descriptions to regionalize the climate: arctic, prairie, maritime, boreal, northern, Great Lakes — St. Lawrence, Pacific — West Coast, Atlantic East Coasts, Cordillera, dry interior, temperate, etc. In this book the climates of Canada are described separately for the provinces and territories rather than according to a climatic type. Highlighted in each account are the important controls that shape the area's climate, and those features that are most characteristic of its climate, e.g., sunshine in Saskatchewan, polar night in Yukon, and fog in Newfoundland. They are not chapters written for provincial travels ads, resort brochures and Chambers of Commerce. On the contrary, extremes and severe weather events are emphasized, which gives the reader the false impression that killer storms, torrential rains, crippling snows, numbing cold, and searing heat dominate an area's climate. Although it is true that each of the provinces and the two territories has climatic extremes and each its usual share of weather excesses and anomalies, the climate is usually tranquil and enjoyable. Indeed, every place has its time under the sun!

For convenience, the data tables, at least one for each province and territory, are listed together near the end of the book.

CANADIAN CITIES RELATIVE TO EURASIAN LOCATIONS

●Pond Inlet

●Whitehorse

●Winnipeg

●Calgary

●Saskatoon

●St. John's

●Edmonton

●Yellowknife

●Victoria

●Vancouver

●Toronto

CANADIAN AND WORLD WEATHER RECORDS

highest maximum air temperature	45.0°C Midale and Yellow Grass, Sask. July 5, 1937	58.0°C Al'azizyah, Libya Sept. 13, 1922
lowest minimum ai r temperature	-63.0°C Snag ,Y.T. February 3, 1947	-89.6°C Vostok, Antarctica July 21, 1983
coldest month	-47.9°C Eureka, N.W.T. February 1979	
highest sea-level pressure	107.96 kPa Dawson, Y.T. February 2, 1989	108.38 kPa Agata Siberia, USSR December 31, 1986
lowest sea-level pressure	94.02 kPa St. Anthony, Newfoundland, January 20, 1977	85 kPa in eye of Hurricane Gilbert September 14, 1988
greatest precipitation in 24 hours	489.2 mm Ucluelet Brynnor Mines, B.C. October 6, 1967	1869.9 mm Cilaos, La Réunion Island March 15, 1952
greatest precipitation in one month	2235.5 mm Swanson Bay, B.C. November 1917	9300 mm Cherrapunji, India July 1861
greatest precipitation in one year	8122.6 mm Henderson Lake, B.C. 1931	26 461.2 mm Cherrapunji, India August 1860—July1861
greatest average annual precipitation	6655 mm Henderson Lake, B.C.	11 684 mm Mt. Waialeaie, Kauai, Hawaii
least annual precipitation	12.7 mm Arctic Bay, N.W.T. 1949	0.0 Arica, Chile—no rain for 14 years
highest average annual number of thunderstorm days	34 days London, Ont.	322 days Bogor, Indonesia

Canada's Climate in Global Perspective

Halifax
Saint John

Few nations can match the climatic diversity of Canada. Canadian climates can be benign, invigorating, or downright inhospitable—all at the same time. Considering the amount of geography Canada covers, such diversity is not surprising. Canada is the largest country in the Western Hemisphere. It is bathed by three oceans and contains the vast interior waters the Great Lakes and Hudson Bay, a veritable sea cutting deep into the centre of the continent. On a global basis, Canada's area is second only to that of the USSR and is greater than that of either China or the United States.

None of Canada's climates has the genial qualities of the tropics or the dryness of the great deserts, the horrific winters of Siberia, the oppressive humidity of the rain forests, or the monotony of the equatorial region, although echoes of other climates around the world can be found, if only because many of the same large-scale controlling factors occur elsewhere.

The fiftieth parallel comes within 400 km of the Maritimes Provinces and passes through the other seven. Following that line east across the Atlantic, one would expect to find twins of Canadian climates in such European cities as Paris, Frankfurt, Prague, Cracow, and Kiev. However, the absence of a massive north-south mountain barrier in western Europe means that these cities are much more exposed to the moderating influence of an ocean than their Canadian counterparts are. To find comparable climates in Europe, it is necessary to travel 600 to 1000 km farther north to Scandinavia and the northern USSR.

In Asia, on the other hand, the fiftieth parallel cuts through locations in southern Siberia, northern Mongolia, northeastern China, and northern Japan which have climates that resemble either the dry continental Prairies or the more humid southeastern regions of Canada. Farther north, the Canadian Arctic shares many climatic characteristics with the Soviet Arctic, although Siberian winters are generally harsher than any in Canada.

A Climate of Moderate Extremes

Canada's first official weather observation was taken at Toronto in early 1840. A century and a half later, with more than 8000 different sites sampled, the Canadian climate archive contains nearly 3 billion observations of temperature, snowfall, wind speed, and dozens of other elements. Surprisingly, in this wealth of data there are no world weather records. In spite of the reputed ferocity of our climate, it has always been colder, snowier, windier, wetter, and hotter somewhere else in the world.

Newfoundland's Salty Climate

AVERAGE ANNUAL NUMBER OF DAYS WITH SOME FOG/SUNSHINE

● Fog
● Sunshine

ST. JOHN'S

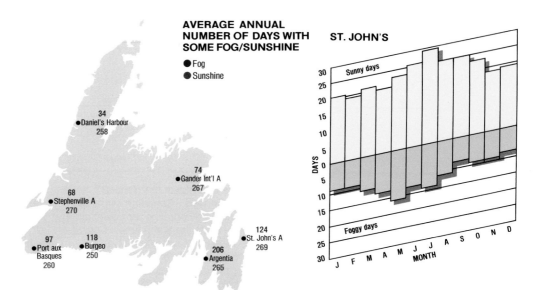

- 34 ●Daniel's Harbour 258
- 74 ●Gander Int'l A 267
- 68 ●Stephenville A 270
- 97 ●Port aux Basques 260
- 118 ●Burgeo 250
- 206 ●Argentia 265
- 124 ●St. John's A 269

A land renowned for its independence and diversity, Newfoundland is home to a fascinating array of climates and weather. Its geography explains many of the unique features of the province's climate. The island covers 5 1/2 degrees of latitude, about the same as the Great Lakes. Its southern extremity lies close to the forty-seventh parallel, approximately the same latitude as Seattle and Paris. It covers an area of 108 860 km², with elevations ranging from sea-level to above 800 m. There are few physical barriers to protect Newfoundland from weather systems sweeping across it. Its situation on the eastern side of North America favours strong seasonal contrasts in the visiting air masses.

Climatically, Newfoundland is the most maritime of the Atlantic Provinces, and this is evident in all seasons, but especially in spring and summer, which are quite cool by Canadian standards.

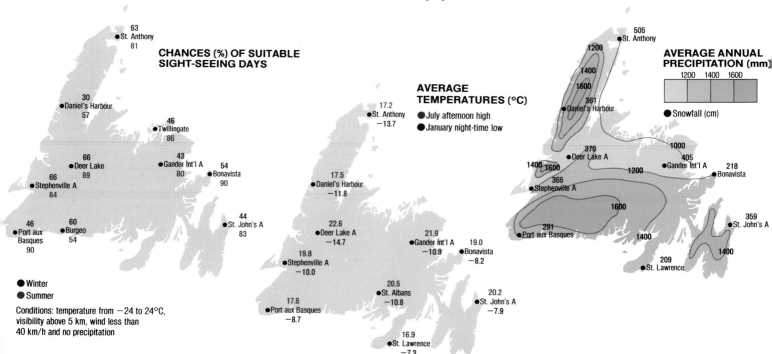

CHANCES (%) OF SUITABLE SIGHT-SEEING DAYS

- 63 ●St. Anthony 81
- 30 ●Daniel's Harbour 57
- 46 ●Twillingate 86
- 66 ●Deer Lake 89
- 43 ●Gander Int'l A 80
- 54 ●Bonavista 90
- 66 ●Stephenville A 84
- 46 ●Port aux Basques 90
- 60 ●Burgeo 54
- 44 ●St. John's A 83

● Winter
● Summer

Conditions: temperature from −24 to 24°C, visibility above 5 km, wind less than 40 km/h and no precipitation

AVERAGE TEMPERATURES (°C)

● July afternoon high
● January night-time low

- 17.2 ●St. Anthony −13.7
- 17.5 ●Daniel's Harbour −11.8
- 22.6 ●Deer Lake A −14.7
- 21.9 ●Gander Int'l A −10.9
- 19.0 ●Bonavista −8.2
- 19.8 ●Stephenville A −10.0
- 20.5 ●St. Albans −10.8
- 20.2 ●St. John's A −7.9
- 17.6 ●Port aux Basques −8.7
- 16.9 ●St. Lawrence −7.3

AVERAGE ANNUAL PRECIPITATION (mm)

1200 1400 1600

● Snowfall (cm)

- 505 ●St. Anthony
- 1200
- 1400
- 1600
- 361 ●Daniel's Harbour
- 370 ●Deer Lake A
- 1000
- 405 ●Gander Int'l A
- 218 ●Bonavista
- 1400 1600
- 1200
- 365 ●Stephenville A
- 1600
- 291 ●Port aux Basques
- 1400
- 359 ●St. John's A
- 1400
- 209 ●St. Lawrence

To Know Newfoundland is to Know the Sea

It is said Newfoundlanders live on, by, and from the sea. No place is more than 100 km from the ocean, and therefore every part of the island is subject to the year-round modifying influences of the encircling cold waters. Surface water temperatures on the Atlantic side range from summer highs of 11 to 13°C inshore and 8 to 11°C offshore to winter lows of -1°C inshore and +2°C offshore. Sea temperatures on the Gulf side are warmer than the Atlantic side by 1 to 3°C. The open sea keeps winter air temperatures a little higher and summer temperatures slightly lower on the coast than at places inland. The marine climate means generally more changeable weather, ample precipitation in a variety of forms (sometimes all at once), higher humidity, lower visibility, more cloud, less sunshine, and stronger winds than a continental climate.

Storm Fury

Ample amounts of low cloud, heavy precipitation, and strong wind over Newfoundland are evidence of the number of storms that pass over and near the island. Indeed, many of the storms that cross North America during the year from west to east, or develop and intensify off the East Coast of the United States, pass near the island while they move out to the North Atlantic. The result is that Newfoundland has a deserved reputation as one of the stormiest parts of the continent. It also has some of the most variable weather anywhere. At all times of the year Newfoundland is near one of the principal storm tracks. The severity and frequency of storms is greatest between November and March, although they may occur at any time of the year.

Winter cyclones are fast-moving storms (up to 80 km/h) that bring abundant and varied precipitation. They pose a serious

---

ST. JOHN'S - CANADA'S WEATHER CHAMPION

Of all the major Canadian cities, St. John's is the foggiest (124 days, next to Halifax's 122), snowiest (359 cm, next to Québec City's 343), wettest (1514 mm, next to Halifax's 1491), windiest (24.3 km/h average speed, next to Regina's 20.7), and cloudiest (1497 hours of sunshine, next to Charlottetown's 1818 hours). It also has more days with freezing rain and wet weather than any other city. But the natives are proud of their climate, calling it character-building and invigorating. And they boast that their city happens to have one of the mildest winters in Canada (third mildest city next to Victoria and Vancouver). Perhaps Townies also happen to appreciate a fine weather day more than the rest of Canadians.

NEWFOUNDLAND

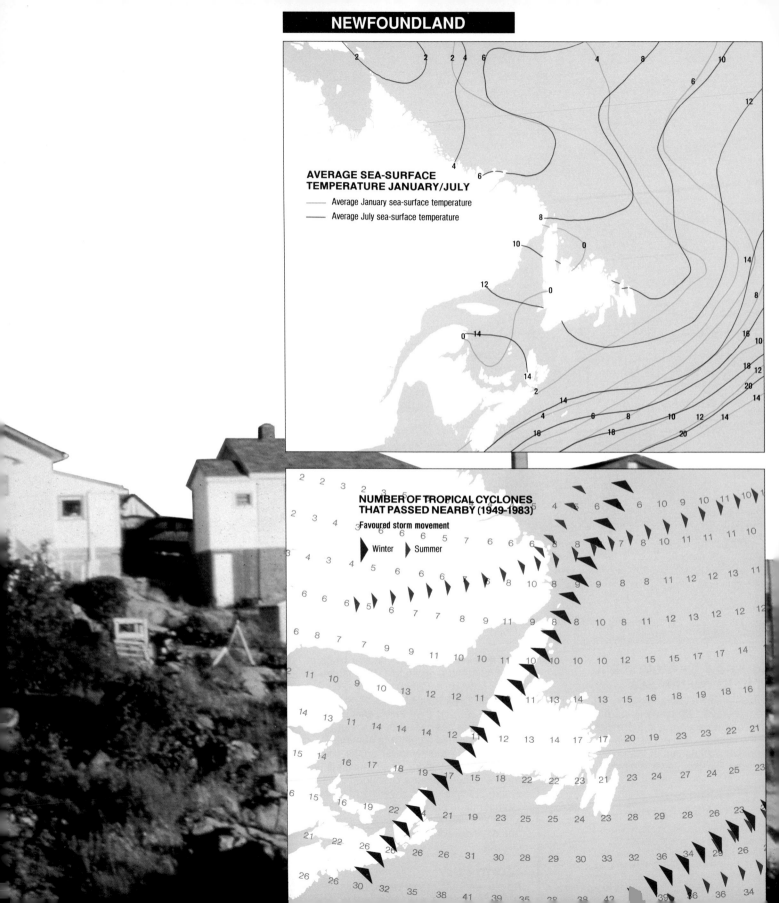

AVERAGE SEA-SURFACE TEMPERATURE JANUARY/JULY

— Average January sea-surface temperature
— Average July sea-surface temperature

NUMBER OF TROPICAL CYCLONES THAT PASSED NEARBY (1949-1983)

Favoured storm movement

◄ Winter   ◄ Summer

threat to fishermen, commercial shipping, and offshore oil and gas exploration activities. Winds often mount to gale and sometimes hurricane force. Hardly a winter goes by without at least three or four East Coast gales.

Occasionally, throughout the year, mature cyclones are prevented from moving out of the region by an upper atmosphere block. The resulting cool, cloudy, and rainy weather associated with the system may persist for a week or more.

During the summer and early fall, Newfoundland weather is typically less stormy. However, in the fall, tropical storms spawned near the equator and developed in the Caribbean may bring windy, wet weather while they pass by the island before dying or redeveloping in the North Atlantic. Over the past thirty-five years, an average of one tropical storm per year has passed within 300 km of Newfoundland. One of the most notorious of these was the "Independence Hurricane" that struck eastern Newfoundland on September 9, 1775. About 4000 sailors, mostly from the British Isles, were reported to have been drowned. On September 5, 1978, another violent storm, Hurricane Ella, passed south of Cape Race. Her winds exceeded 220 km/h. At St. John's, 45 mm of rain fell and winds reached 115 km/h.

Big Blow
Newfoundland as a whole has the strongest winds of any province, with most stations recording average annual wind speeds greater than 20 km/h. Generally, coastal stations have stronger winds than inland stations, valleys have lighter winds than elevated terrain, and winter is decidedly windier than summer.

Bonavista on the East Coast is the windiest location, with an average annual wind speed of 28 km/h, St. Albans, in the sheltered Bay d'Espoir on the south coast, is the least windy location, with an average yearly speed of 11.5 km/h.

Winds are predominantly from the west year-round, but variations are common both from location to location and from month to month. Prevailing wind directions are west in winter and west-southwest in summer. Calm or light and variable conditions occur about 2 to 3% of the time along the coast but more than 10% of the time at inland stations.

Ocean Air
Temperatures in Newfoundland tend to be midway between those of Winnipeg and Vancouver. This is true of both average temperatures and extremes. Newfoundland's temperature range (the difference between the average temperatures of the warmest and coldest months) is 20°C. In comparison, Winnipeg's is 38.9° and Vancouver's is 14.8°.

Winter temperatures in Newfoundland show the day-to-day variability that is characteristic of a stormy maritime climate. Incursions of moist, mild Atlantic air are frequent. There is also a noticeable difference between inland and coastal temperatures. In the interior, winter tempera-tures average between -6°C and -10°C, whereas on the southeast coast, where the moderating influence of the ocean is greatest, the winter average is between -2°C and -4°C. The lowest Newfoundland temperature on record is -41.1°C, set at Woodale Bishop's Falls on February 4, 1975.

Spring comes rather slowly and is short. Until late May, night-time averages in the interior are under 4°C, and in many valley locations there is a 90% probability of frost on any night until the first week of June.

Summer is also short and cool. The glacial Labrador Current holds July average temperatures in coastal areas around 14°C, but inland averages may climb over 16°C. Sunny summer days in Newfoundland, however, are among the most delightful anywhere in Canada. With afternoon highs in the low twenties, they are warm enough to be comfortable and yet cool enough to permit vigorous activity. Summer 1987 was especially pleasant across central Newfoundland. Record high sunshine, scanty rainfall, and seasonable temperatures pleased most residents and tourists. The highest temperature ever recorded on the island is 36.7°, occuring at Botwood, northeast of Grand Falls, on August 22, 1976.

The frost-free period varies widely. Typical growing-season lengths are 150 days on the south coast, 125 on the Avalon Peninsula, and under 100 days in the interior. In some locations, where the landscape causes cold air to collect at night, frost-free durations may be only 70 days on average. The growing season in interior Newfoundland tends to start around June 8 and to end by September 15. Generally, frost is not a limiting factor for the growing of root vegetables, such as potatoes and turnips, which are the most important of the few crops grown in Newfoundland.

Wet and Wild

With the exception of some stations on the north coast, yearly precipitation totals exceed 1000 mm everywhere across the island. The greatest amounts occur inland along the south coast between Port aux Basques and St. Albans, which lies in the path of Atlantic storms, and over the Long Range Mountains, which lie in the lee of the Gulf of St. Lawrence. The south coast has annual totals exceeding 1650 mm, making this region the wettest in eastern Canada.

Approximately three-quarters of the total precipitation falls over the island as rain and one-quarter as snow . Although precipitation is well distributed throughout the year, it is heaviest in fall, with November being the wettest month. Spring is usually the driest time of the year. Newfoundland can experience the occasional drought; however, its intensity and duration are less than in central and western Canada. The summer of 1987 had rainfalls between 50 and 70% of normal, causing problems for farmers and keeping the forest fire hazard index high all season.

Snowfall dominates winter precipitation. It is heavy, with normal amounts exceeding 300 cm at most places in the province. Along the south coast, however, snowfall totals are in the 200-300 cm range, less than elsewhere because much of the storm precipitation falls as rain. Some of the heaviest snowfalls occur at higher elevations along the west coast, when cold air outbreaks crossing the Gulf of St. Lawrence meet the Long Range Mountains. Snowfalls here total over 400 cm a year and are heaviest in January before the Gulf has frozen over. Corner Brook, with an average of 410 cm of snow a year, and Gander with 405 are two of the snowiest locations on the island. St. John's, with 360 cm, is the snowiest major city in Canada.

Silver Thaw

Freezing rainstorms are a major hazard in many parts of Canada, but they are nowhere more frequent than in Newfoundland where they are known as silver thaws. Some of these storms can be very severe, causing extensive damage and bringing transportation and other essential services to a standstill. One of the worst ever experienced on the island struck St. John's on the evening of April 11, 1984, and continued intermittently until the 14th. Jackets of ice as much as 15 cm thick formed on overhead wires. The interruption of power supplies left 200 000 persons in the Avalon Peninsula without heat and light for days, causing a run on kerosene heaters at retail stores.

The area between St. John's and Gander is especially prone to prolonged periods of

freezing precipitation that last for several hours or intermittently for two days or more, disrupting everyday activities and damaging trees and property. Freezing rain or freezing drizzle occurs an average of 150 hours each winter, with March being the worst month.

More "Smoke" than Sun

Newfoundland is not noted for its sunshine. The total number of hours of bright sunshine for the island is usually less than 1600 hours a year, which is below Summerside's average of 1959 hours, Calgary's 2314 hours, and the Canadian average of 1925 hours. The summer months are the sunniest, with an average of 187 hours of

sunshine a month, about 42% of the total possible. The least sunshine is experienced in December, when the average daily duration is about 2 hours.

The waters off the Avalon Peninsula and over the Grand Banks are among the foggiest in the world. The fogs, sometimes known as "sea smoke," develop when warm, humid air from the south strikes the cold, sometimes ice-infested, waters of the Labrador Current.These fogs may occur in all seasons, but, on average, they are most frequent in the spring and early summer when the contrast between sea and air temperatures is greatest, anywhere between 5°C and 15°C. Argentia has 206 days of fog, Belle Isle and Cape Race over 160, and St. Lawrence on the Burin Peninsula 147.

Surprisingly, the fogs are often accompanied by strong winds. Normally, winds can be expected to disperse fog, but here the fog is frequently so dense and widespread that the winds have little clearing effect. The resulting conditions can be hazardous for shipping and for drilling rigs, especially when icebergs are present.

Floes and Wintry Ghosts

During the first half of each year the waters off Newfoundland may become choked with ice floes and icebergs. The severity of ice varies considerably, depending on the strength and direction of the wind and the coldness of the air. In a normal year, ice enters the Strait of Belle Isle by the beginning of January. The ice edge usually reaches Notre Dame Bay by the end of the month and Cape Freels by the middle of February. On the west side, Labrador ice moves into the Gulf of St. Lawrence through the Strait of Belle Isle, but the vast majority of the ice is formed within the Gulf itself and the estuary. The ice edge reaches its maximum southern extent in March, filling the innumerable bays and coves and effectively retarding the advent of spring.

In April the rate of melting overtakes the southward ice drift and the pack slowly retreats. Normally by mid-month navigation through the Strait of Belle Isle is possible.By mid-June the median ice edge retreats to the mid-Labrador coast. In extreme years ice may linger south of Belle Isle after Canada Day.

Each year an average of 250 icebergs drift along in the cold waters of the Labrador Current onto the Grand Banks. These majestic, wintry ghosts worry mariners more than pack ice, chasing drill platforms off site or barricading fishermen in the many bays and harbours. Icebergs have been counted and tracked since the sinking of the *Titanic* in 1912. Although 250 is an average number, the yearly extremes have ranged from none in 1966 to 2202 in 1984.

GRAND BANKS CLIMATE DIGEST

	Jan.	April	July	Oct.
Temperature °C				
Air - average	0.1	1.6	11.6	8.5
High Extreme	17.8	17.5	24.5	21.5
Low Extreme	-11.5	-7.0	-11.0	-1.0
Sea Surface	0.7	0.7	10.0	8.6
Relative Humidity %				
Average	87	91	96	86
Wind				
Speed (knots)	25	20	18	20
Extreme Speed (knots)	95	87	86	86
Prevailing Direction	W	SW	SW	NW
Waves				
Height (m)	2.5	1.9	1.2	1.9
Max. Height (m)	14.1	12.0	6.2	11.5
Period (s)	6	6	5	6
Max. Period	21	16	14	18
Percentage of Observations With				
Precipitation	36	20	11	24
Visibility > 3 km and wind < 25 knots	46	47	32	57
Visibility < 1 km	9	22	52	9

More Arctic Than Atlantic

With its rugged rock and, bleak fiorded coast, and largely unexplored interior, Labrador is one of the most unspoiled parts of Canada. Its climate may be described as somewhere between inhospitable and invigorating.

The climate of Labrador is more Arctic than Atlantic. Because it is on the eastern side of the continent, it experiences strong seasonal contrasts in the characteristics and movement of air masses. The predominant flow is off the land. The rugged Torngat Mountains in the north, with peaks above 1500 m, and the Mealy Mountains in the south, with peaks about 1200 m, confine the moderating influence of the Atlantic Ocean to the rocky islands and near shore.

The limitation of the ocean's influence, however, is not a serious disadvantage, because in this region its effect on the climate is generally unpleasant. The Labrador sea is infested with floating pack ice and icebergs for eight months of the year. The masses of ice keep sea temperatures below 4°C. An east wind off the Labrador Current is a cool wind in summer, often with light rain or drizzle. In winter, when the Atlantic air is relatively mild, the accompanying weather includes cloud and frequent snowflurries. Whenever easterly winds bring very moist air from the Atlantic, widespread fog occurs.

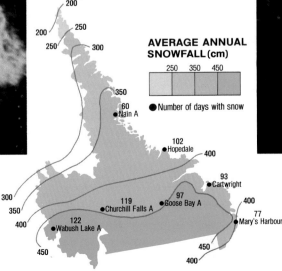

AVERAGE ANNUAL SNOWFALL (cm)

250 350 450

● Number of days with snow

200
200 250
250 300
350
60
● Nain A
102
● Hopedale 400
93
● Cartwright
119 97
● Churchill Falls A ● Goose Bay A 400
300 77
350 ● Mary's Harbour
122
● Wabush Lake A
400
450 450
400

Temperature

Winters are very cold, with typical daytime temperatures for January between -10 and -15°C, colder than Newfoundland and more like the frigidity of the southern Prairies. An occasional incursion of Atlantic air will warm up the winter. The summer season is brief and cool along the coast because of the cold Labrador Current. July average temperatures are from 8 to 10°C along the coast but are 3 to 5°C warmer in the interior. The pleasantness of the summer day along the coast is often determined by the wind direction—westerly winds bring clear, mild continental air, whereas easterlies, blowing off the Labrador Current, bring cold, cloudy, and moist weather.

Precipitation

Precipitation is heaviest in the south and decreases northwards. On the whole it is much lighter than in Newfoundland, although amounts can vary considerably from year to year.

Southern Labrador is not unlike the moist northern shores of Newfoundland, with 1000 mm as a typical yearly fall of precipitation. About 45% of this occurs as snow. Over much of Labrador 800 mm is a more typical amount, with about half of it snow. In summer, rainfall is quite reliable, with seasonal totals seldom less than 175 mm in the north and 275 mm in the south.

Snowfall is heavy, with Churchill Falls in the interior having 481 cm, making it one of the snowiest places in Canada. Goose Bay has a mean snowfall of 445 cm. In the south, Cartwright averages 440 cm, and in the north Nain is typical with 424 cm. The ground is snow-covered for eight months in the far north and for six months in the south.

NEWFOUNDLAND AND LABRADOR WEATHER SUPERLATIVES

	What	Where	When
Temperature			
highest	41.7°C	Northwest River (Labrador)	Aug. 11, 1914
lowest	-51.1°C	Esker 2 (Labrador)	Feb. 17,1973
Greatest			
precipitation	2253 mm	Pools Cove	1983
snowfall	893 cm	Woody Point	1977-78
number of fog days	230	Argentia Airport	1966
wet period	266 days	Corner Brook	1949
number of blowing snow days	76	Belle Isle	1958-59
sunshine	1806 hours	Goose Bay (Labrador)	1974
wind (maximum hourly)	138 km/h	Belle Isle	Feb. 2, 1976
frost-free period	225 days	Burin	1926

METEOROLOGICAL MOMENT

Early Monday morning on February 15, 1982, the giant drilling rig *Ocean Ranger* capsized and sank on the Grand Banks, 300 km east of St. John's. The entire 84-man crew perished in the violent winter seas, marking the worst Canadian marine disaster in decades. It was the world's second worst catastrophe in offshore drilling history, next to the North Sea tragedy on March 22, 1980, when 123 died.

The deadly *Ocean Ranger* storm began as a weak disturbance in the Gulf of Mexico on Friday the 12th. By Saturday evening it was centred south of Nova Scotia, collecting its strength and moving toward the Avalon Peninsula of Newfoundland. By Sunday afternoon the storm was located near St. John's and had a central pressure of 95.4 kPa. For most of the 14th the rig was battered by hurricane-force winds that reached 168 km/h and by waves as high as a five-storey building. The rig soon developed a list of 12 to 15 degrees on the port side. In the early hours of the 15th the order went out to abandon the *Ocean Ranger* for the stormy North Atlantic. Soon after daybreak the world's largest semi-submersible offshore rig slipped beneath the wild seas off Newfoundland.

Search and rescue crews battled poor visibility in freezing rain and snow, as well as freezing spray, turbulent seas, and buffeting winds in an attempt to locate survivors. But there were none.

The storm also contributed to the sinking of the Soviet container ship *Mekhanik Tarasov* with a loss of 33 lives about 120 km east of the site where the *Ocean Ranger* sank.

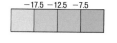
●Goose Bay A

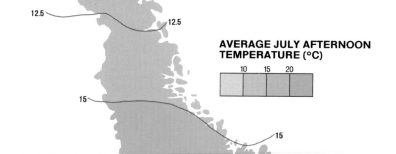

AVERAGE JANUARY AFTERNOON TEMPERATURE (°C)

-17.5	-12.5	-7.5

AVERAGE JULY AFTERNOON TEMPERATURE (°C)

10	15	20

●Goose Bay A

A Moist and Temperate Isle

Canada's Garden of the Gulf is surrounded by sea, lying between the Northumberland Strait on the south and west and the Gulf of St. Lawrence on the north and east. With the oceanic influence, the onset of the seasons is delayed several weeks. Winters are rambunctious but, on the whole, milder than in most parts of Canada. Spring is late and cool. Summer is modest and breezy. As for fall, well, Islanders will tell you they favour this season most of all, except when it involves the occasional brush with a dying Atlantic hurricane.

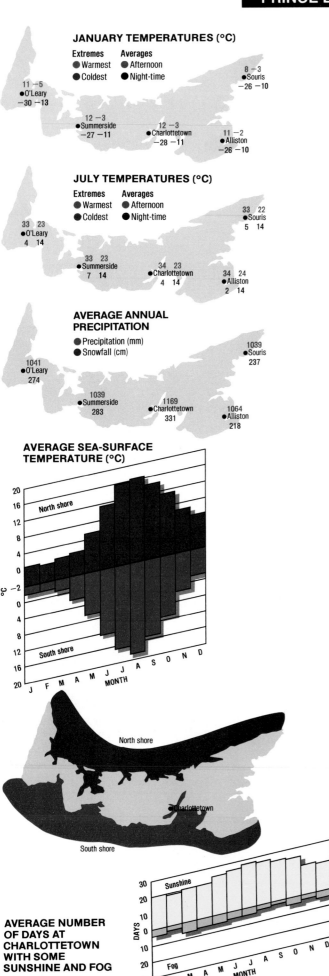

JANUARY TEMPERATURES (°C)

Extremes
- ● Warmest
- ● Coldest

Averages
- ● Afternoon
- ● Night-time

O'Leary 11 −5 / −30 −13

Souris 8 −3 / −26 −10

Summerside 12 −3 / −27 −11

Charlottetown 12 −3 / −28 −11

Alliston 11 −2 / −26 −10

JULY TEMPERATURES (°C)

Extremes
- ● Warmest
- ● Coldest

Averages
- ● Afternoon
- ● Night-time

O'Leary 33 23 / 4 14

Souris 33 22 / 5 14

Summerside 33 23 / 7 14

Charlottetown 34 23 / 4 14

Alliston 34 24 / 2 14

AVERAGE ANNUAL PRECIPITATION

- ● Precipitation (mm)
- ● Snowfall (cm)

O'Leary 1041 / 274

Souris 1039 / 237

Summerside 1039 / 283

Charlottetown 1169 / 331

Alliston 1064 / 218

AVERAGE SEA-SURFACE TEMPERATURE (°C)

North shore

South shore

MONTH: J F M A M J J A S O N D

North shore
Charlottetown
South shore

AVERAGE NUMBER OF DAYS AT CHARLOTTETOWN WITH SOME SUNSHINE AND FOG

Sunshine
DAYS
Fog
MONTH: J F M A M J J A S O N D

Oceanic Heat Pump

The island's gently rolling terrain, which generally does not rise any higher than 50 m above sea-level, presents no formidable obstacles to the weather systems that traverse the region. The sea is a much more important climate control, with no community farther than 20 km from the sea-shore. The seas about the Island function like a gigantic heat pump, drawing warmth from the waters in the fall and early winter and cooling the air for a greater part of the spring and summer seasons. From January to early April, when the gulf and straits become ice covered, the heat pump shuts down and the Island becomes as continental as the interior of New Brunswick.

Drift ice is often found around Island seas as late as the latter part of May, causing difficulty for fishermen and retarding the arrival of spring. The ice cover is by no means solid throughout this period, but is broken up by tides, strong winds, and occasional thaws. Just as weather profoundly affects the extent and duration of ice cover, so the relative amount of ice and water affects the weather, influencing the occurrence of fog, cloud, snow, blowing snow, and extreme temperatures.

Weather Mosaic

Prince Edward Island has some of the most variable day-to-day weather experienced anywhere in the country. A specific set of weather conditions seldom lasts for long. The Island is affected by a mishmash of weather systems, bringing polar, maritime, continental, and tropical air from the Arctic, Pacific, and Atlantic oceans and from the Gulf of Mexico. In summer the Island is visited most often by continental air from the west. Occasionally, warm tropical air is drawn in from the south. Winters are dominated by cold continental air masses, but storms originating in the North Pacific or the Gulf of Mexico frequently pass through, making winters stormy and unpredictable. The influence of moist Atlantic air often produces warm periods during the winter and cool weather during the summer.

Cloudy Not Foggy

Cloud, mist, haze, and a bit of fog conspire to keep sunshine totals in P.E.I. below the national average, but not too far below. Its 1905 hours of bright sunshine are fairly close to the national average of 1925 hours. However, less than half the total possible sunshine occurs in every month and only 20% of the possible amount occurs in December.

The Island is relatively free of fog year-

PRINCE EDWARD ISLAND WEATHER SUPERLATIVES			
	What	Where	When
Temperature			
highest	36.7°C	Charlottetown (Agriculture Canada station)	Aug. 19, 1935
lowest	-37.2°C	Kilmahumaig	Jan. 26, 1884
Greatest ...			
precipitation	1597 mm	Alliston (Agriculture Canada station)	1979
snowfall	539 cm	Charlottetown Airport	1971-72
number of fog days	59	Charlottetown Airport	1967
wet period	206 days	Charlottetown CDA (Agriculture Canada station)	1983
number of blowing snow days	40	Charlottetown Airport	1971-72
sunshine	2132 hours	O'Leary	1978
wind (maximun hourly)	93 km/h	Borden	Feb. 2, 1976
frost-free period	199 days	Charlottetown	1931

round, since neighbouring provinces help to shelter P.E.I. from fog-bearing southerly winds off the cool Bay of Fundy and the Atlantic Ocean. Even so, spring and summer are the foggiest seasons. At this time the ocean waters chill the warm, moist air masses moving over them, causing huge banks of persistent fog. Cold season fog results when the air is made moist and warm by contact with open water. Days with fog at Summerside number 37 a year, compared with 101 at Halifax (Shearwater) and 106 at Saint John.

Never Hot but Can Be Cold
Summers are pleasantly cool. The July average temperature is about 18.5°C, with daily highs normally in the low to mid-twenties. Maximum temperatures exceed 30°C only once or twice a year and have never exceeded 37.8°C (100°F) at any official weather station on P.E.I.

Winter temperatures reflect the presence of sea ice, becoming colder later in the winter after the surrounding sea-water freezes. Bitterly cold temperatures below -18°C occur on only four or five nights in any winter, and thaws can occur in all the winter months. Periods free from frost extend over 130 days between late May and early October, a duration which compares favourably with that north of Lake Ontario and is about a month longer than the growing season on the southern Prairies.

Wet and Breezy
Precipitation is reliable and ample year-round. Annual totals average just under 1000 mm over the southeastern tip around Montague, increasing to 1100 mm or more in the centre of the Island at Charlottetown and New Glasgow. Monthly totals are greatest in the late fall and early winter, exceeding 100 mm monthly from October to January at Charlottetown, due to the more frequent and more intense storm activity at that time of the year. Wet days number 130 to 160 a year on the Island, with snow days accounting for 30% of them. Combined with the long growing season, the generous rainfall is responsible for the Island's verdant landscape and good farming.

Measurable snowfalls are frequent over the long winter season from November to April. The Island is one of the snowiest parts of Canada. Charlottetown's 330.6 cm makes it the third snowiest city next to St. John's and Québec. However, because the snow comes and goes throughout the winter, there is an appreciable cover for only about three-quarters of the snow season, which lasts from late November to April 20, a period of about 145 days.

Heavy snowfalls of 12 cm or more average five a year, with one or two of them leaving as much as 25 cm of snow. The biggest snowfalls are usually associated with large, slow-moving weather systems and may last for up to 4 days.

Winds on the island are on average stronger than at most Maritime locations. The strongest winds occur in the coldest months, when stormy weather prevails. The lightest winds are summer breezes. Generally, winds from October to April are from the west or northwest; in summer, winds from the south and southwest predominate.

Wide Open to Storm Fury
The Island seldom experiences violent local storms in the form of tornadoes, severe thunderstorms, and hailstorms, although waterspouts, the aquatic cousins of tornadoes, are somewhat more in evidence. Thunderstorm days number between 9 and 12 a year on average considerably less than in other parts of southern Canada, and thunderstorms are certainly not of the same

METEOROLOGICAL MOMENT
Long-time residents could not remember a more vicious winter storm than the one that blew across the Gulf of St. Lawrence for a week at the end of February 1982. The onslaught began on the 21st and continued on the 22nd with 25 to 45 cm snowfalls and 100 km/h winds in zero visibility. An additional 10 cm of snow fell during the week, but gale-force winds would not let up, blowing freshly ploughed snow back into place. Wind chills between -30 and -40°C were common during the siege, making venturing outdoors foolish if not dangerous. Snowdrifts of 5-7 m high towered over buildings and forced snowploughs off the Trans-Canada Highway. Under a stat-e of emergency, government helicopters were used to airlift food and fuel and provide medical services to marooned residents. Three snowploughs and two freight trains were stuck in drifts for up to a week. It took two weeks before roads were made passable and life returned to something resembling normal. Few who experienced it will forget the week-long blizzard of February 1982.

intensity as those experienced elsewhere. There are occasional exceptions, however. On July 17, 1980, heavy rains from a violent thunderstorm flooded parts of Charlottetown and blacked out the city for many hours.

Although local storms are rarely severe, the Island is vulnerable to the destructive forces of much more powerful Atlantic storms. These bring very high tides (storm surges), strong winds, and heavy rains. About once every summer and early fall, dissipating hurricanes tracking along the Atlantic coast expend their energy and remaining rainfalls over the Island. One of the most drenching and damaging storms of this kind occurred on September 22, 1942. Charlottetown recorded 163.8 mm of rain, the greatest daily total ever recorded for any P.E.I. station.

Winter storms pose a serious challenge. Packing a variety of weather conditions from hurricane force winds to heavy precipitation in all forms, they can pass rapidly through the region or stall and batter the province for days. Winds associated with these storms on occasion exceed 100 km/h. When such storms occur at high tide, storm surges become a problem. When the centres of the storms remain to the south of the Island, precipitation reaches the Gulf and the Island in the form of snow. If the low centre passes to the north of the Island, the snow will change to freezing rain and rain. Freezing rain and snowfall may combine to paralyse the Island for days, at great cost to communications and transportation. During an average year, about 40 hours of freezing rain or drizzle fall over Prince Edward Island, coating everything in sight.

Blizzards, which unleash the threat of snow, wind, and cold, strike the province with paralysing force about two or three times a winter.

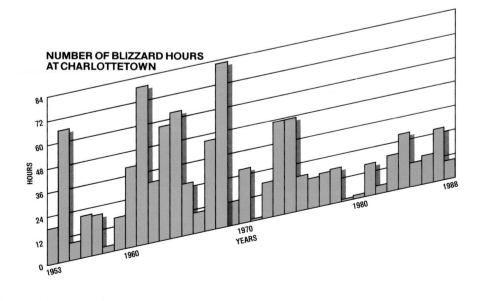

NUMBER OF BLIZZARD HOURS AT CHARLOTTETOWN

Nova Scotia The Stormiest Region of Canada

Described on provincial vehicle licence plates as Canada's Ocean Playground, Nova Scotia owes everything to the sea, especially its climate. Robust winters, reluctant springs, fresh summers, and lingering falls; reliable precipitation and lavish snowfalls; misty sunlight, thick fog, and expansive sea ice — all of these, and many more, are a part of Nova Scotia's maritime climate. The influence of the sea is not surprising. The province is virtually a peninsula surrounded by seas: the Gulf of St. Lawrence to the north, the Bay of Fundy to the west, and the Atlantic Ocean to the south and east.

By the Sea
Atlantic and Fundy waters are relatively cold (8 to 12°C), and they help to keep the air temperature over southwestern Nova Scotia on the cool side in spring and summer. In

January, when their temperature is between 0 and 4°C, these same waters moderate the harshness of winter. Farther offshore to the east, southeast, and south are the comparatively warm 16°C waters of the Gulf Stream. It's warmth, especially from August through October, is credited with prolonging fall — the season many Nova Scotians consider to be the best of the year.

Nova Scotia's north coast is exposed to Gulf waters which, in late August, have a maximum surface temperature of 18°C. The contrast between air and water temperatures is enough to create onshore sea breezes and to hold back the onset of fall for a few weeks. In January, however, Gulf of St. Lawrence and Northumberland Strait waters become ice-covered, effectively cutting off any marine influence for the next three months.

The influence of the sea is felt in other ways. Ice conditions in the Gulf and, on occasion, in the Bay of Fundy retard the arrival of spring. Cool summer seas also help to stabilize overriding air masses, thus

suppressing local storm development. In addition, the merging of contrasting ocean currents — the warm Gulf Stream and the cold Labrador Current — produces a great deal of sea fog that often moves far inland.

The effects of latitude and relief on climate are not as important in Nova Scotia as elsewhere. The highest mountains are the Cape Breton Highlands, part of the Appalachian mountain chain, but nowhere are they higher than 530 m. Although this is not high enough to block the movement of air masses, it is sufficient to divert them and cause them to move in a direction parallel to the mountains. Locally, the Highlands help to wring additional moisture from passing weather systems and to cool temperatures by 1 or 2°C.

Warmest In the East
The southwest coast around Cape Sable is frost-free for over half the year, longer than any other place in Atlantic Canada and comparable to localities along the shores of Lake Erie. Most agricultural areas experi-

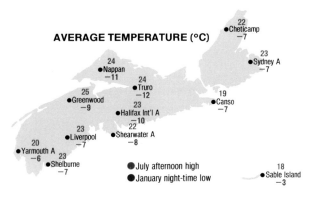

AVERAGE FROST-FREE PERIOD (DAYS)

140
100
80
120
120
120
140
120
100
80
160
• Halifax Int'l A
140
140
160
160
60
140
80
160

80 120 160

200

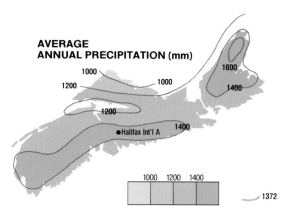

AVERAGE TEMPERATURE (°C)

22 • Cheticamp −7
23 • Sydney A −7
24 • Nappan −11
24 • Truro −12
19 • Canso −7
25 • Greenwood −9
23 • Halifax Int'l A −10
22 • Shearwater A −8
23 • Liverpool −7
20 • Yarmouth A −6
23 • Shelburne −7
18 • Sable Island −3

● July afternoon high
● January night-time low

AVERAGE ANNUAL PRECIPITATION (mm)

1000
1200
1000
1600
1400
1200
1200
• Halifax Int'l A
1400

1000 1200 1400

1372

AVERAGE ANNUAL SNOWFALL (cm)

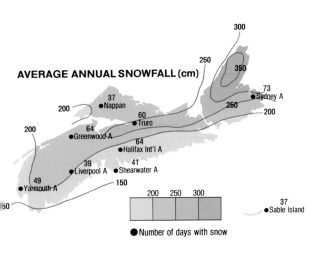

300
250
350
73 • Sydney A
250
200
37 • Nappan
200
60 • Truro
200
64 • Greenwood A
64 • Halifax Int'l A
39 • Liverpool A
41 • Shearwater A
49 • Yarmouth A
150
50

200 250 300

37 • Sable Island

● Number of days with snow

78

ence a period without frost for 120 to 130 days between late May and early October, which is the effective growing season for most crops. The Annapolis Valley has up to 140 frost-free days, but the higher highlands on Cape Breton Island have less than 100 days.

Winter temperatures are moderate along the coast. Yarmouth's average January temperature of -2.7°C is the highest of any mainland station in the Maritimes. Inland, January means are between -4 and -6°C. The most significant aspect of winter is the marked day-to-day variation caused by the alternation of Arctic and maritime air.

Summers are relatively cool in Nova Scotia. Afternoon summer temperatures reach 25°C in the interior, but along the coast are frequently 4 to 6°C cooler. At night the ocean remains a cooling source, keeping minimum temperatures along the coast about 2 to 3°C below those inland. Halifax's July mean of 17.4°C and Yarmouth's 16.3°C compare closely with Vancouver's 17.3°C but are somewhat cooler than Toronto's 20.6°C.

Mainly Moist

Nova Scotia is wettest over the highlands of Cape Breton Island, where over 1600 mm of precipitation fall in an average year. The southern coast experiences almost as much, with totals of 1500 mm. By contrast, the north shore along the Northumberland Strait has less than 1000 mm a year.

Precipitation is slightly greater in the late fall and early winter because of the more frequent and intense storm activity. In most years there is a good supply of rain during the growing period. However, drought is not unknown in Nova Scotia. A prolonged warm, dry, and sunny spring in 1986 contributed to the worst forest fire outbreak in the province's history. The previous summer, several months of below normal precipitation dried up wells and streams, and water levels did not begin to recover until Hurricane Gloria brought heavy rains in late September.

On average, only about 15% of Nova Scotia's total annual precipitation originates as snow. An

exception is northern Cape Breton Island, where the snow fraction is closer to 30%. Snowfall is relatively light near the warm Atlantic shore and near the entrance to the Bay of Fundy, where less than 150 cm may fall in one winter. Here, copious rain and freezing rain make up for the scanty snowfalls. Inland, the yearly snowfall increases to 250 cm. As a rule, elevated areas receive the greatest snowfall and have the longest snowcover season. Both the Cobequid Mountains and the Cape Breton Highlands receive in excess of 300 cm in an average snow year. These elevated areas also experience "sea-effect" snowfalls in the wake of winter storms. Heavy local snowfalls are also produced by winds blowing off the open waters of the Gulf of St. Lawrence and the Bay of Fundy.

The snowcover season, that is, the period when there is at least 2.5 cm of snow

	Halifax (City)	Vancouver (UBC)
Annual Precipitation (mm)	1282	1258
Rainfall (mm)	1065	1203
Snowfall (cm)	217	55
% of Precipitation October to March	57	74
Yearly Days with Precipitation	153	167
Wettest Month	November	December

Both Halifax and Vancouver get about 1300 mm of precipitation a year, yet, as the table shows, there are some interesting differences. Perhaps the most interesting is that Vancouver has a distinct winter rainy season, whereas Halifax experiences a more even spread of precipitation throughout the year.

on the ground, varies considerably. Usually its duration extends from about 110 days a year along the southern coast to 140 days inland and in areas adjacent to the frozen seas. In coastal areas the snowcover may come and go. At Halifax, for instance, there is only a 50% chance that there will be snow on the ground for Christmas.

Some of the more notable provincial snowfall extremes are — greatest snowfall in one season: 653 cm at Cheticamp in 1964-65; greatest in one month: 224 cm, also at Cheticamp, March 1961; and greatest in one day: 69 cm at Yarmouth on March 19, 1885.

Misty Sunshine

Halifax's reputation as a foggy, misty city is well deserved. Each year there is an average of 122 days with fog at the International Airport and 101 days at Shearwater, on the Dartmouth side of the harbour, although on most days fog persists for less than 12 hours. The period from mid-spring to early summer is the foggiest time. Bands of thick, cool fog lie off the coast, produced

where the chilled air above the Labrador Current mixes with warm, moisture-laden air moving onshore from the Gulf Stream. With onshore winds these banks of fog move far inland. Sea fog often affects the headlands by day, moving inland and up the bays and inlets at night. At other times of the year fog is much more transient and local in nature.

Besides Halifax, other foggy places are Yarmouth (118 days), Canso (115 days) and Sydney (80 days). No part of Nova Scotia is fog-free, although some places inland from the Minas Basin have no greater fog frequency than Toronto. Nova Scotia's most persistent spell of fog occurred during Canada's centennial in 1967 at Yarmouth, when over the 92 days of summer, 85 had an occurrence of 1 or more hours with fog.

Because of the extensive fogs, as well as mists, low cloud, and smog, sunshine amounts throughout the province are usu-

NOVA SCOTIA METEOROLOGICAL MOMENT

Remembered in weather chronicles as the Great Nova Scotia Cyclone, a calamitous hurricane swept over Cape Breton Island on August 25, 1873. The storm was unusual at that time in having travelled so far to the east after leaving the tropics. Its destructive power was also extraordinary. Ravages of the storm included 1200 vessels, 500 lives, 900 buildings, and an untold number of bridges, wharves, and dykes. Property losses were conservatively estimated at $3 1/2 million, an amount equivalent to $70 million in 1990. At the height of the storm, gale-force winds were reported at Halifax, Sydney, and Truro. Also noted in the weather records for these stations were observations of an intense thunderstorm and heavy rainfalls of 50 mm or more. The Sydney weather observer remarked that this was the worst gale since 1810.

Losses were high partly because the interruption of telegraph service between Toronto and Halifax prevented storm warnings from getting through. Of significance to Canadian meteorology, this storm, perhaps more than any other event, convinced officials of the need for an improved Canadian storm warning system.

ally less than half the total possible. Apart from foggy Sable Island, sunshine totals range from 1700 to 1969 hours a year. July is the sunniest month inland, and August is the sunniest along the coast. Sunless days (days with less than 5 minutes of bright sunshine) amount to between 75 and 90 a year, with a marked seasonal high from November to February. Sunny days, on which less than 70% of the sky is covered with cloud in the early afternoon, amount to between 130 to 160, with a peak from July through October.

Stormy and Changeable

Storms frequently pass close to the Atlantic coast of Nova Scotia and cross the southern part of Newfoundland, producing highly changeable and generally stormy weather. Without doubt this region has more storms over the year than any other region of Canada. Winter storms are especially devastating, with occasional loss of life and extensive damage to property. Packing a variety of weather conditions from hurricane-force winds to heavy precipitation, they can pass rapidly through or stall and batter the region for days. On occasion, the winds associated with these nor'easters, as they are called, exceed 150 km/h, and peak wave heights can be as high as 14 m. At high tide, these winds can cause storm surges of more than a metre. Other conditions associated with these storms include freezing spray, reduced visibility in snow, rain, or fog, and numbing wind chills,

SABLE ISLAND—GRAVEYARD OF THE ATLANTIC

Situated 300 km east-southeast of Halifax are the crescent-shaped, shifting sand dunes of Sable Island. Home of the wild Sable Island ponies with the long, flowing manes and tails, it has also been a temporary home for shipwrecked sailors, enroute convicts, and pirates brought there inadvertently by the legendary gales that blow around the island. Sunken ships, the victims of these storms, litter the surrounding ocean floor, giving the island its reputation as "the graveyard of the Atlantic."

Sable's climate is influenced mainly by the sea. Ocean waters maintain winter air near freezing and keep summer temperatures below what would be considered room temperature in milder parts of the country. The average annual range is only 18.6°C. This compares with Halifax International Airport's 24.3°C and Winnipeg's 38.9°C. Recorded temperatures have never gone lower than -19.4°C (January 31, 1920) or higher than 27.8°C (August 27, 1951). February is the coldest month and August is the warmest.

The ocean is also a prime source of moisture for cloud and precipitation. The yearly rain and snow total averages 1372 mm, of which only 9% is snowfall. The number of wet days almost equals the number of dry days. Even the driest months, July and September, receive 92 mm of precipitation each, while the wettest, December and January, both receive 145 mm. Snowfall and freezing precipitation totals and occurrences, however, are less than on the mainland.

Since the island is in the path of travelling frontal storms year-round and tropical cyclones in the summer and fall, most of its precipitation comes from large-scale storms. Thunderstorms are infrequent, with only 11 thunderstorm days a year, but hurricanes and tropical storms, such as George in 1950, Helene in 1958, and Evelyn in 1977, are notorious for the heavy rainfall they bring.

Lying near the track of these storms, Sable is sometimes battered by winds much stronger than those blowing on the mainland, and the treeless terrain offers little protection. Strong gales, waves, currents, and sea swell are constantly reshaping the island, eroding shoreline cliffs at one end and elongating the crescent shape at the other.

Although Sable is exposed to winds from all directions, they are mostly from the southwest during spring and summer and shift to a westerly direction during fall and winter. Wind speed records are impressive. The highest hourly speed observed is 130 km/h. The highest gust officially recorded is 174 km/h, although there is an unofficial report of a gust speed close to 190 km/h.

Adding to Sable's inhospitable weather label is its reputation as the foggiest place in the Maritimes. On average, 127 days a year have at least one hour of fog. In comparison, Summerside has fog on 37 days a year. Unfortunately, the sunniest and driest time — the summer — is also the time with the most fog. July has 22 fog days on average. In 1967 fog was present on 30 of 31 July days, and June of the same year recorded the longest duration of fog, 126 hours — more than five whole days of it.

The climate capsule that emerges for this bleak but fascinating island is stormy times, mild in winter, cold in summer, and humid and windy all year-round.

JOHN DEARY

SHIPPING HAZARDS NEAR SABLE ISLAND

PER CENT OF TIME

Visibility below 1.6 km

Winds above 50 km/h and/or freezing precipitation

J F M A M J J A S O N D
MONTH

especially in the storm's wake.

In late summer and fall the remnants of a hurricane or tropical storm are felt at least once a year in Nova Scotia. The most memorable of these dying storms was Hurricane Beth on August 15-16, 1971. At Halifax, Beth brought 296 mm of rain, more than the deluge from Hurricane Hazel over Toronto and enough to wash away several bridges, damage buildings, and flood farmland.

Other severe weather phenomena include ice storms and blizzards. Each year one or two 25-cm snowfalls occur in Nova Scotia and newspaper headlines across Canada announce another paralysing Atlantic snowstorm. When combined with strong winds, they cause enormous inconvenience and, at times, property damage and loss of life. A record snowstorm struck Halifax on February 2 - 3, 1960, producing a total of over 75 cm of snow in the downtown area and over 96 cm in the suburbs, most of it within 24 hours.

On the other hand, Nova Scotia is not known for spectacular displays of thunder and lightning. Thunderstorms occur on about 10 days of the year, about half the number that occurs in northern and central New Brunswick. Tornadoes have been recorded but are rare. One such tornado, accompanied by heavy hail and lightning, struck White Point Beach near Liverpool on January 30, 1954, but most weather watchers consider it a freakish event. Reports of waterspouts over nearshore waters are received yearly.

variation that occurs in both wind direction and speed results from the characteristics of natural and man-made obstructions, topography, and surface cover. Along the coast, an onshore sea breeze circulation often sets up, particularly during a warm, sunny afternoon in the spring or early summer.

NOVA SCOTIA WEATHER SUPERLATIVES

	What	Where	When
Temperature			
highest	38.3°C	Collegeville	Aug. 19, 1935
lowest	-41.1°C	Upper Stewiacke	Jan. 31, 1920
Greatest ...			
precipitation	2360 mm	Wreck Cove Brook	1983
snowfall	653 cm	Cheticamp	1964-65
number of fog days	152	Sable Island	1966
wet period	230 days	St. Paul Island	1936
number of blowing snow days	35	Sydney Airport	1964-65
sunshine	2246 hours	Halifax (Shearwater)	1978
wind (maximum hourly)	135 km/h	Halifax Citadel	Jan. 19, 1980
frost-free period	267 days	Sable Island	1953

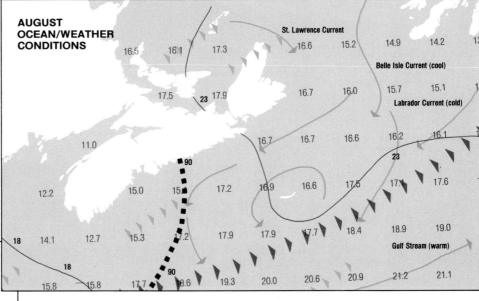

LEGEND

- ■ ■ ■ ■ Per cent of time wave heights less than 2 m
- ▷ ▷ ▷ ▷ Prevailing wind direction
- —— Average wind speed (km/h)
- – – – – Per cent of time wave heights above 6 m
- 4.6 14.2 Sea-surface temperature (°C)
- → Ocean current
- ◣ ◣ ◣ Winter storm tracks
- ▶ ▶ ▶ Summer storm tracks

Winds

Winds blow predominantly from the south or southwest in the summer with an average speed of about 10 to 15 km/h. In the coldest months the predominant direction is from the west and northwest with an average speed of 22 km/h.

The wind at any given location is often quite different from the wind conditions which prevail even a short distance away. The

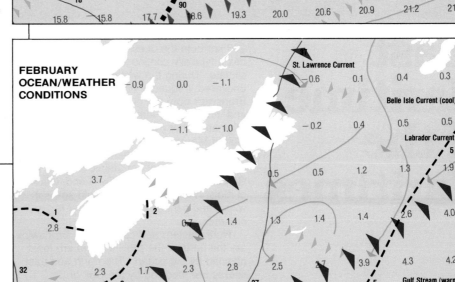

ENVIRONMENT CANADA

Least Maritime of the Maritimes

No part of New Brunswick lies more than 200 km from the ocean, yet the province has a typically continental flavour to its climate. During the winter, cold air, largely unaltered, frequently flows across New Brunswick from the centre of North America, and most storms affecting the province originate either over the North Pacific or the Gulf of Mexico. In summer, the predominant air mass is warm continental, with occasional incursions of hot, humid air from the Gulf of Mexico. On the other hand, influxes of moist Atlantic air produce mild spells in winter and periods of cool weather in summer.

Northwestern and central New Brunswick are the parts of the province that are least affected by the ocean. The north and east shores are modified by the Gulf of St. Lawrence, but less so than their proximity might suggest. The extent of the sea-effect depends to a large degree on the wind direction, with onshore winds causing the most moderation. The cold Gulf waters retard warming of the air in spring, keep the summer maximum temperature low, and provide a slight warming of the air in fall— provided the winds are off the water. Near the Bay of Fundy, continental air masses are modified by the ocean. Coastal locations such as St. Andrews and Saint John experience moist Atlantic air most of the year, producing mild periods during the winter and cool weather the rest of the year.

Yet, in spite of their by-the-sea location, New Brunswick coastal areas are not good examples of true marine climates. Continental climates, compared with true maritime climates, have an earlier spring and shorter fall, wider fluctuations of temperature from day to day and from season to season, and more snowfall but less total precipitation. Near the coast there is a blending of these continental and maritime influences, with the temperature moderat-

ing effect of the ocean increasing the closer we get to the coast.

There are two key indicators that tell us whether a climate is influenced more by the ocean or by the land in northern latitudes. The first is the season of maximum precipitation: in marine climates it is the cold season. The second is the percentage of the total annual precipitation that falls as snow: less than 25% for marine areas. Saint John's climate qualifies as marine on both counts.

The Land

The southern landscape is characterized by hills sloping down to tidal marshes at the edge of the Bay of Fundy, whereas the eastern and central portions of the province consist of rolling hills cut by river valleys. The highlands in the northwest are an extension of the Appalachian mountain chain. Mount Carleton, the highest point in the Maritimes, is found here. At 820 m

above sea level, however, its elevation is still 240 m lower than Calgary's.

Apart from local effects and an influence on the direction of air motion, topography has little influence on New Brunswick's climate.

Temperature

A PROVINCE OF CONTRASTS

January is the coldest month in New Brunswick and July is the warmest. Along the Fundy coast, average daytime highs vary between 20 and 22°C in the summer. These temperatures are reached fairly early in the day, by 11 a.m. or noon, before the sea breeze sets in and causes them to drop sharply. The sea breeze is not as prevalent along the Gulf coast, however, partly because the Gulf tides are not as pronounced as the Bay of Fundy's. Also, as the warm season progresses, the shallow waters of the Gulf heat up faster and reach higher temperatures than those of Fundy, thus diminishing the difference between air and water temperatures. Since it is precisely this difference which creates the sea breeze, the strength of the breezes off the Gulf are decreased.

As we move inland, the moderating effect of the ocean diminishes, and the interior regularly experiences temperatures of 25°C and higher during the summer. Along the Gulf of St. Lawrence, afternoon temperatures regularly reach 24°C, except on those days when the cool sea breeze prevails. On several occasions, extremes in New Brunswick have exceeded 37.8°C (100°F). The highest temperature ever recorded in the province is 39.4°C, set at Nepisiguit Falls, Rexton, and Woodstock on August 18 and 19, 1935.

In winter, temperatures decrease noticeably from south to north. The interior,which has elevations above 600 m and is more directly in the path of continental air masses, experiences very cold winters. At Edmundston, the January mean temperature is -12.2°C. As we move south, however, this coldness is gradually tempered by the effects of latitude and, to a greater extent, the sea. Along the southeastern shores, the January mean is around -7.5°C.

Frigid temperatures are not infre-

quent, with most settlements in the northwest reporting extreme low temperatures of -30 to -35°C every winter. The all-time provincial low is -47.2°C, recorded at Sisson Dam, a weather station near Plaster Rock, N.B., on February 1, 1955. This is also the low-est temperature recorded anywhere in the Maritimes.

Perhaps the most significant feature of a New Brunswick winter is the marked variability in temperature from day to day. This is a product of the highly contrasting and fast-moving weather systems which traverse the province every two or three days.

The usual differences between coastal and inland localities are also seen in the length of the frost-free season. Along the Fundy shore 140 to 160 frost-free days occur on average, whereas in the central highlands of the Miramichi there are less than 90 days. A four-month frost-free season extends along the Gulf coast, and along the south shore of the Bay of Chaleur the growing season is extended by another week.

Precipitation

THE SNOWIEST BUT NOT THE WETTEST

Generally across the province, as the winter snowfall increases from place to place, the total precipitation for the year decreases. Cold winter temperatures and stormy northeasterlies combine to make New Brunswick the snowiest of the three Maritime provinces. Northwestern New Brunswick generally receives between 300 and 400 cm of snow annually, for about 33% of its annual total precipitation (rain and snow combined). On the other hand, the eastern and southern sections of the province receive 200 to 300 cm of snow, less than 20% of their annual total precipitation. Winter storms frequently bring rain to the Fundy coast and snow to the interior.

At Saint John the season's average snowfall of 293 cm occurs on 59 days

CONTINENTAL AND MARINE CLIMATES

	Continental Climate (Edmundston)	Modified Marine Climate (Saint John)	Marine Climate (Sable Island)
mean temperature of coldest month (°C)	-12.2	-6.7	-1.0
temperature range (°C) (summer maximum - winter minimum)	42	33	24
total snowfall (cm)	364	293	126
total precipitation (mm)	1121	1444	1372
wet days	163	163	173
snow days	65	59	37
wettest month/ season	Aug./summer	Dec./winter	Jan./winter
degree-days below 18°C	5271	4483	3813
frost-free period (days)	112	171	197
last spring frost	May 28	May 2	Apr. 29
first fall frost	Sept. 18	Oct. 21	Nov. 13

between the middle of November and the middle of April. Although snowfalls are often deep, the snowcover comes and goes quickly. This pattern may repeat itself three or more times during the winter. In the northwest, however, the snowcover season lasts 160 days and is both reliable and persistent.

Spring and early summer are notably dry over New Brunswick, but there is ample water during the growing season. The interior highlands record about 1200 mm of rainfall a year, with the heaviest amounts falling during the summer months, a pattern characteristic of a continental-type climate. On the other hand, the southern shoreline receives a like amount, but with a slight maximum in fall and early winter. Elsewhere, 1000 mm is a representative yearly amount.

Stormy Weather

On average, thunderstorms occur between 10 to 20 days a year at any place in New Brunswick, and this is higher than over the remainder of Atlantic Canada. Normally only one thunderstorm a year is severe enough to produce hail. Tornadoes are rare but occur more often in New Brunswick, especially in the northwest, than in Nova Scotia. Even though tornado sightings are unusual, evidence of trees blown down by strong winds is not. When tornadoes do strike, they are usually rated as minor. This was not the case, however, on August 6, 1879, when a tornado ravaged the village of Buctouche, demolishing everything in its path. Seven people were killed, 10 were injured and 25 families were left homeless.

Freezing precipitation is a hazard on about a dozen or more days a year in New Brunswick. Hours of freezing rain or drizzle range from 34 hours a year at Fredericton to 59 hours at Moncton.

Storms associated with low pressure areas can occur at any time of the year but tend to be more severe and frequent in winter. These winter storms pack strong winds with snow often changing to rain and back again to snow. One of the most memorable storms of this kind in recent history struck eastern New Brunswick on January 4, 1986, walloping Moncton with 110 km/h winds and 67 cm of snow in 24 hours.

In the summer and fall, the main storm track passes through the Bay of Fundy and northeast through the Strait of Belle Isle. Weakened tropical storms and hurricanes affect the southern portion of the province with at least one heavy rainstorm every one or two years.

Winds blow predominantly from the west and northwest in the cold months and from the south and southwest in the warm

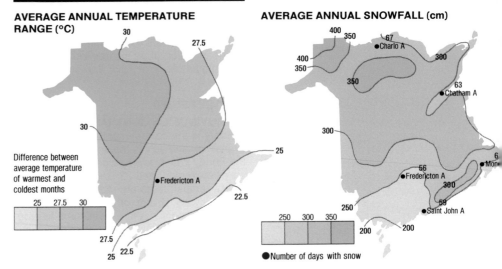

AVERAGE ANNUAL TEMPERATURE RANGE (°C)

Difference between average temperature of warmest and coldest months

25 27.5 30

AVERAGE ANNUAL SNOWFALL (cm)

250 300 350

● Number of days with snow

months. Wind speeds average 15 to 20 km/h in winter and 12 to 15 km/h in summer. With some exceptions, winds tend to be lighter farther away from the sea coast. Along the coast, sea breezes are frequent, especially in the early afternoon on warm, sunny days under light regional winds. Local exposure is very significant. Open, treeless localities and the tops and slopes of ridges and hills are often very windy. Also, local topography causes funnelling of the wind in many circumstances.

Windless conditions occur more fre-

quently inland, anywhere between 5 to 12% of the time depending on local exposure. Coastal sites have fewer calms, with windless conditions occurring only 1 to 5% of the time. Strong winds above 50 km/h blow mainly from the west. Severe winds approaching hurricane force occur an hour or two each year along the coast, but are rarely measured at distances inland.

Sunny Maritimes

New Brunswick has some of the sunniest places in Atlantic Canada: Chatham is the only station in the region to record an

DOUGLAS SMALL

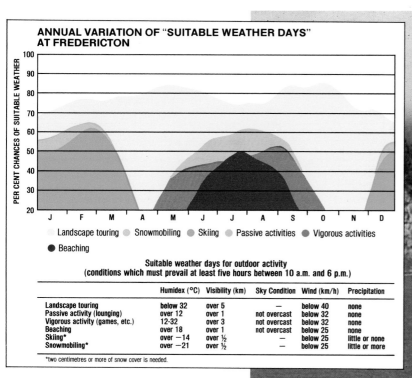

ANNUAL VARIATION OF "SUITABLE WEATHER DAYS" AT FREDERICTON

(Chart: PER CENT CHANCES OF SUITABLE WEATHER, 20–100, months J F M A M J J A S O N D)

Legend:
- Landscape touring
- Snowmobiling
- Skiing
- Passive activities
- Vigorous activities
- Beaching

Suitable weather days for outdoor activity
(conditions which must prevail at least five hours between 10 a.m. and 6 p.m.)

	Humidex (°C)	Visibility (km)	Sky Condition	Wind (km/h)	Precipitation
Landscape touring	below 32	over 5	—	below 40	none
Passive activity (lounging)	over 12	over 1	not overcast	below 32	none
Vigorous activity (games, etc.)	12-32	over 3	not overcast	below 32	none
Beaching	over 18	over 1	not overcast	below 25	none
Skiing*	over −14	over ½	—	below 25	little or none
Snowmobiling*	over −21	over ½	—	below 25	little or more

*two centimetres or more of snow cover is needed.

average of 2000 hours a year. Perhaps the most surprising statistic is that Saint John in December reports nearly 100 hours of bright sunshine on average, more than any other station in eastern Canada! On the other hand, in July it reports one of the lowest sunshine totals in Canada. Across New Brunswick in an average year, sunless days number 75 and sunny days number between 140 and 160.

Fundy Fog

The well-mixed Atlantic waters off the Bay of Fundy are among the foggiest areas of the world, although not as notoriously foggy as the Grand Banks of Newfoundland. The season with the greatest contrast in temperature between sea surface and overriding air produces the greatest fog. At Saint John fog occurs on more than one quarter of the days of the year and 36% of the time in July. Sea fog is much more prevalent during the night and early morning than during the day. At Saint John early morning fog occurs on 60% of the fog days; by 2 p.m. the fog frequency drops to 18%.

Elsewhere across New Brunswick, the fall is the foggiest season, with occurrences on 4 or 5 days each month, but overall conditions are not unusually foggy. For example, in an average year, foggy days number 44 at Chatham, 24 at Campbellton, and 60 at Moncton; in comparison, Toronto has 35, London 50, Regina 29 and Vancouver 45.

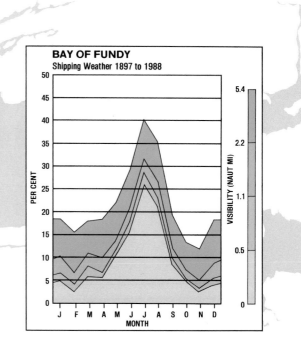

BAY OF FUNDY
Shipping Weather 1897 to 1988

(Chart: PER CENT 0–50 on left axis, VISIBILITY (NAUT MI) 0, 0.5, 1.1, 2.2, 5.4 on right axis; MONTH J F M A M J J A S O N D)

A MARINER'S TALL TALE
"Sometimes out on the Bay of Fundy when the fog comes in thick you can sit on the boat's rail and lean your back up again it. So that's pretty thick fog out there; but you gotta be careful 'cause if the fog lifts quick you'll fall overboard."

River Flooding

The Saint John River has a history of flooding dating back to the arrival of the first European settlers. Within this century there have been at least six floods which have caused damage in excess of one million dollars. The worst flooding occurred in spring 1973 along the lower Saint John River in the Fredericton area, resulting in economic costs of $11.9 million ($ 40 million in 1990 dollars). A flood such as this is likely to occur only once in 70 years.

Conditions in late April 1973 were ripe for serious flooding. The winter snowpack was high, equivalent to 300 mm of water in some places. On April 27, two days after rivers crested at above normal levels, a storm moved into northern and central New Brunswick, bringing mild temperatures and a 75-mm rainfall. Rivers rose rapidly, in many cases to record levels, and peaked on April 29 and 30.

The total economic cost of the flood was almost $12 million. The most seriously affected area was the flood plain of the lower Saint John River in the Fredericton area and the farmlands downstream of Fredericton. Among the impacts on agriculture were a shortened growing season, physical damage to buildings and machinery, and reduced yields from crops and livestock. Even more serious was the loss of fertile soil.

At the peak of the flood, basements and ground floors of homes were flooded, trailers were waterlogged, and cottages were ravaged by turbulent waters and floating debris. In Fredericton 260 homes and 193 businesses were damaged. The Lord Beaverbrook Hotel was closed for a few weeks as a result of the flooding.

NEW BRUNSWICK WEATHER SUPERLATIVES

	What	Where	When
Temperature			
highest	39.4°C	Woodstock	Aug. 18, 1935
		Nepisiquit Falls	Aug. 18, 1935
		Rexton	Aug. 19, 1935
lowest	-47.2°C	Sisson Dam	Feb. 1, 1955
Greatest ...			
precipitation	2150 mm	Alma	1979
snowfall	704 cm	Tide Head	1954-55
number of fog days	127	Saint John Airport	1971
wet period	208 days	Tide Head	1954
number of blowing snow days	24	Saint John Airport	1974-75
sunshine	2226 hours	Chatham	1978
wind (maximum hourly)	137 km/h	Miscou Island	Oct. 21, 1968
frost-free period	204 days	Grand Manan Island	1902

ENVIRONMENT CANADA

NEW BRUNSWICK METEOROLOGICAL MOMENT

February 2 is usually the day when the groundhog checks his shadow, but for Saint Johners it has another significance. On that day in 1976 the city was racked by one of the fiercest storms ever to hit the Bay of Fundy in this century. At the height of the storm, winds were clocked at 188 km/h and waves reached heights of 12 m with swells as high as 10 m. Stronger winds have occurred, but they did not last as long as they did during the infamous Groundhog Day Storm.

Locals claimed that no storm since the legendary Saxby Gale of 1869 had caused such extensive damage. Tens of millions of dollars in losses were counted, including damaged docks and buildings, battered craft and mobile homes, and severed hydro poles and trees. Utilities were out and transportation slowed for a week. Everything in sight became coated with sea salt, when wind-driven spray was carried far inland. Strong winds blowing with the tide resulted in exceptionally high water levels that eroded huge chunks of coastline. An unusual phenomenon, never seen before, was the disappearance of many sea-birds, which never returned. Adding to the destruction were many broken water lines, frozen by bitterly cold Arctic air arriving on the heels of the windstorm.

Although the storm affected parts of central New Brunswick and all of Prince Edward Island, southern New Brunswick, especially Saint John, and western Nova Scotia were the hardest hit.

Immense
Land...

Immense
Climate

Quebec is Canada's largest province, 1.5 million square kilometres, larger than the combined area of nine countries, of western Europe. Because of its enormous latitudinal extent, Quebec exhibits a wide range of climates. In the northern and central part of the province, winters are long and cold and summers are the opposite. This region has a distinctly subarctic landscape — sparse, barren, and lightly treed in the far north and on the tops of the Shield's ridges, and more heavily forested or with a lichen woodland in the river valleys and lake areas across the central and southern portions of the Shield. The north shore of the Gulf of St. Lawrence has more of a marine climate, with wet, stormy winters and cool summers, heavy snow, and frequent fog. Sea ice clogs the waters of the estuary and Gulf for up to four months a year.

Climatically, southern Quebec is typically Canadian — brisk winters with a snow-thaws with rain in winter. Mild air from the west predominates year-round in the south, and northerly air that is cool in summer and cold in winter is most frequent in the north. Following the freeze-up of Hudson Bay in late December, cold air masses with dry conditions and clear skies move freely across the entire province. On occasion, air from the northern Atlantic enters the province via Labrador. These air masses are cool and humid in summer but mild, cloudy, and wet in winter. In the north they are responsible for occasional midwinter thaws that produce spectacular amounts of melted snow and turn runways and roads into ribbons of mud.

Winter lows originating in the south or west of the United States usually cross the continent via the Great Lakes and the St. Lawrence corridor. An interesting mix of snow, freezing rain, or rain and periods of mild weather accompanies these lows. In summer, these systems are weak and usually interrupt the sequence of warm,

covered landscape and warm summers with occasional spells of hot, humid weather. It has a continental climate much like that of southern Ontario but with colder winters and more precipitation year-round. The St. Lawrence Valley is the stormy meeting ground of cold, dry air masses from the north and warm, humid air from the south. All regions reveal four distinct seasons and a truly variable climate.

Weather Systems

The cliché, if you don't like the weather just wait a few hours, is certainly apropos of southern Quebec. The day-by-day altera-tion of high and low pressure systems, especially in winter, brings a great variety of weather to the region. Moist air from the Gulf of Mexico brings hot, humid, and unsettled weather in summer and snow and

LOTS OF WHITE STUFF

Each year snow costs the Quebec economy millions of dollars. No major world city has as much snow as Montréal nor spends as much in removing it from the streets—$45 million for 1700 km of streets, not counting the costs of ploughing driveways and parking lots. In a long winter, an incredible 40 million tonnes of snow falls on the city.

But snow is a boon not a bane to Quebec. In fact, an insufficient snowcover can be a disaster. Each year snow returns billions of dollars to the province's economy. Snowmelt means copious water to turn huge hydro-electric power turbines. Tourists and sports enthusiasts flock to Quebec to ski, snowshoe, and drive snowmobiles. Farmers appreciate the insulating quality of the snow and its nourishing meltwater. Foresters build snow roads for winter access. Snow also transforms a drab landscape into a glistening winter wonderland.

Snow is plentiful everywhere in Québec. The St. Lawrence Lowlands and Ottawa Valley receive from 200 to 300 cm in an average year. Downriver from Québec City, and in the Gaspé Peninsula, 400 cm is a typical amount. The heaviest snowfall in eastern Canada, 500 cm or more, occurs north of the Gulf of St. Lawrence in Quebec and Labrador, where moisture-laden air rides up the steep slopes of the north shore. In northern Ungava, the average snowfall is less than 200 cm, but amounts increase at higher elevations and along the Hudson Bay coast, where in the fall the prevailing northwesterly winds deposit heavy snowfalls on the exposed shore until the ice cuts off the supply of moisture in December or later.

Extreme values show a remarkable range at many stations. For example, the greatest and least snowfall in one season at Montréal was 443 cm in 1886-87 and 93 cm in 1979-80. At Québec City, it was 555 cm in 1872-73 and 165 cm in 1948-49. The greatest monthly totals ever recorded for Montréal and Québec City are 159 cm (January 1898) and 209 cm (February 1939) respectively. The province's most memorable single snow event occurred on March 4, 1971, when 47 cm of snow fell on Montréal. High winds whipped the snow into huge drifts and it took more than a week before the city got back to near normal.

humid, hazy weather and cool, dry, clear weather with a spell of showers and unsettled conditions.

Storms that travel northward east of the Appalachian Mountains often affect Quebec. Depending on their track and intensity, they may bring heavy or light snows, a soaking, or a glazing. Another source region for storms affecting Quebec is western and northern Alberta. While these systems cross Hudson Bay and Davis Strait, they draw very warm, humid air into Quebec from the American south.

The Shape of the Land

Quebec has a fairly level landscape. With only 7% of the province lying above 600 m, there are no major mountain barriers capable of blocking traversing weather currents. On the other hand, its topography is effective in wringing out copious amounts of rain and snow from passing systems. Geographically, Quebec has three major relief features: the Laurentian Plateau, the Appalachian Highlands, and the St. Lawrence Lowlands.

The Laurentian Plateau is part of the Canadian Shield, an enormous, rather monotonous expanse of rounded hills with thousands of lakes. The shield covers 80% of the province. Its highest summits rise to over 900 m in the Laurentian Park north of

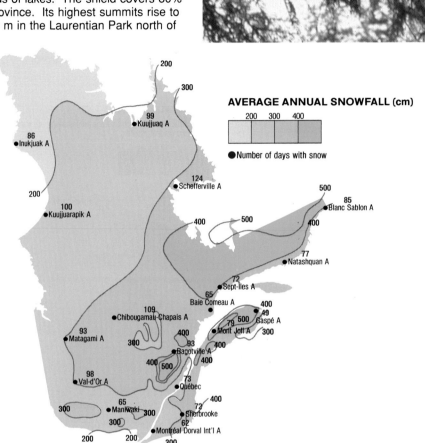

AVERAGE ANNUAL SNOWFALL (cm)

200   300   400

● Number of days with snow

Québec City. Along the north shore of the St. Lawrence, the southern edge of the Laurentians rises abruptly and spectacularly from the river valley. The crest of the escarpment has some of the wettest slopes in Canada, with 1400 to 1600 mm of precipitation yearly.

The Appalachian Highlands extend through the southern Eastern Townships and the Gaspé Peninsula. The region consists of an ancient plateau, rising to about 400 m, with higher ridges and deep valleys that are aligned from southwest to northeast parallel to the regional storm tracks.

The St. Lawrence Lowland is wedged between the Appalachians and the Shield and runs from Québec City to the Ontario border. Level in some parts and rolling in others, it is the most fertile and best developed region of Quebec.

Marine Effects

Quebec has 13 773 km of coastline, most of it bathed by cold Arctic waters. Hudson Bay brings Arctic water as far south as 51°N along the western side of the province. September is the only month without ice in Hudson Bay. With winds off the water, coastal areas receive low cloud, fog, dampness, and cold temperatures. In the fall, when the water is near the freezing point and warmer than the air, extensive cloud and frequent snow flurries cross over the shoreline. In winter an ice-covered Hudson Bay has little modifying effect on the air passing above it.

On the Arctic coast, in Hudson Strait, and in Ungava Bay, ice floes and icebergs are present year-round, chilling the water so that its temperature never exceeds 3°C. However, the ocean never freezes over completely in these areas. The continuous movement from tides, currents, and wind keeps enough water open in winter to generate cloud and precipitation and to moderate the temperature on the coast.

Waters along the Labrador coast also remain icy year-round, reaching a maximum of 4°C by late summer. Prevailing winds are offshore across the continent, but an occasional onshore wind does blow, and the accompanying weather is almost always cold, cloudy, and damp.

The north shore of the Gulf of St. Lawrence has ice from late December until May. A loose thin cover, areas of open water, and ice leads are present all winter. As a result of the marine effect, Quebec's north shore has higher fall and winter temperatures and heavier precipitation than areas away from the Gulf or downwind of the ice cover. In summer, the cooling of warm southerly air by the 10-14° waters of the Gulf can result in cold, misty, and cloudy conditions. In October, when the water temperature averages about 7 to 9°C, the Gulf becomes a heat source for warming the overlying cooler air.

The Great Lakes have little, if any, moderating effect in Quebec.

Six-Month Winters

MON PAYS, CE N'EST PAS UN PAYS, C'EST L'HIVER *Gilles Vigneault*
Quebecers are experienced at winter. The snow season is long, lasting 4 to 5 months in the south and 5 to 8 months through central and northern districts. Winters are also severe. In January average day and night temperatures hover around -5 and -14°C at Montréal and -8 and -17°C at Québec City and in the Eastern Townships. The coast of the Gaspé Peninsula and the gulf islands are milder than this by 4 to 6C. Northern Quebec has corresponding daily readings of -20 and -27°C, and let-ups from

the entrenched Arctic cold are infrequent. In the south, however, there are 3 to 4 days each winter when night-time temperatures remain above freezing.

In southern Quebec, spring arrives suddenly and sometimes not at all. March is almost always wintry, but by April average temperatures are nearly as warm as at Toronto. The thermal climb is rapid until July, when afternoon and night-time temperatures average 26 and 16°C at Montréal and 25 and 13°C at Québec. Farther down

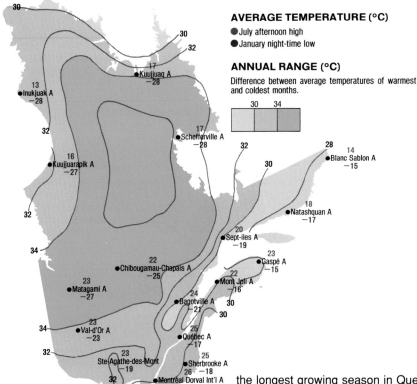

AVERAGE TEMPERATURE (°C)
● July afternoon high
● January night-time low

ANNUAL RANGE (°C)
Difference between average temperatures of warmest and coldest months.

30   34

the St. Lawrence River, where the tempering influence of the sea is felt, midsummer temperatures are 3 to 5°C cooler. On the north shore, average day/night readings decrease from 22°/12°C along the shore to 19° / 9°C on the Laurentian Plateau. Around Hudson Bay and Ungava Bay and in the Labrador Sea, ice floes are still present in July. The coasts have a distinctive subarctic summer, with average temperatures 5°C below those in the centre of the northern land mass.

Warm summer days may occur everywhere in the province but excesses of heat and humidity are rarely a problem. Montréal has an average of only 14 days a year when the weather is hot and muggy for 6- to 12-hour periods and the humidex reaches uncomfortable levels, and only 2 days a year when people would truly suffer.

At Québec City the number of moderate to severe humidex hours is considerably less.

Fall may be as fleeting as spring but almost always is pleasant. Indian summer's hazy warmth and the glorious multicoloured landscape make this the favourite time of year for most.

Frost

Parts of the Ottawa Valley and the Upper St. Lawrence Lowlands around Montréal have

the longest growing season in Quebec. Montréal's frost-free period lasts 157 days, from early May to early October. In the Gaspé Peninsula, coastal sites may have up to 130 frost-free days, but the season is shortened to 100 days and less at higher elevations. Elsewhere in the St. Lawrence Lowlands and along the Gulf's north shore, frost is unlikely for 90 to 120 days. In the north and west this period shortens rapidly, with the length of the season varying sharply from year to year. On the Laurentian Plateau, the frost-free period shows considerable variation with local topography. A valley or depression in which cold air pools on clear, still nights may have fewer than 40 consecutive frost-free days in a year. In nearby areas where air drainage is uninhibited, the frost-free season may last 120 to 130 days.

Mostly Moist

Precipitation in Quebec is reliable, adequate, and remarkably well distributed throughout the year. In the south, frontal showers and thunderstorms in summer, tropical storms in late summer and early fall, and snowstorms in the winter combine to ensure monthly precipitation totals of 60 to 100 mm. Dry or wet spells are not common, yet forest fires occur every year and droughts are still a risk to agriculture. Also, the occurrence of floods and mudslides is a reminder of the impact of excessive moisture.

Precipitation totals rise steadily eastward. Annual precipitation totals in southern Quebec are generally the highest in the province, exceeding 1000 mm at all localities and surpassing 1200 to 1400 mm over the steeply rising slopes along the north shore of the St. Lawrence and the Gulf. Québec receives about 1175 mm, making it one of the wettest cities in Canada. Slightly more of its precipitation occurs in the summer than in the other seasons. Seasonal amounts also show a summer maximum in the St. Lawrence Lowlands, but there is a shift towards fall and winter in coastal regions bordering the Gulf of St. Lawrence.

Annual precipitation is quite heavy in the central interior of the province, averaging 1000 mm at Chibougamau and 800 mm at Nitchequon, but decreases toward the west and north to 400 mm at Inukjuak and 300 mm at Quaqtaq. In both the central and

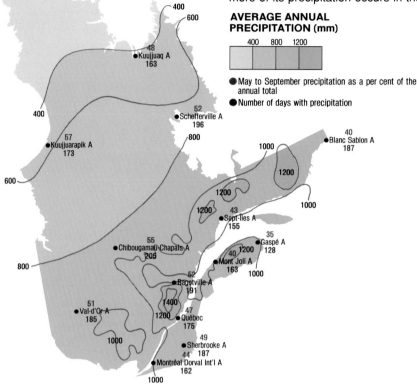

AVERAGE ANNUAL PRECIPITATION (mm)

400 800 1200

● May to September precipitation as a per cent of the annual total
● Number of days with precipitation

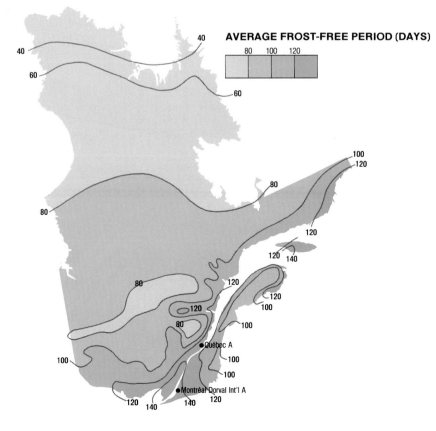

AVERAGE FROST-FREE PERIOD (DAYS)

80 100 120

METEOROLOGICAL MOMENT

The year 1816 is legendary in weather history. It has been called poverty year, the year with the black summer, and the year without a summer. Between May and September, a series of cold waves chilled southern Quebec and New England, causing near-famines in some areas. From June 6 to 10, Québec City was covered by nearly a foot of snow, with drifts reaching "to the axle trees of carriages." Freezing temperatures in June blackened crops and produced thick ice on ponds, trapping and killing large flocks of waterfowl. There were several chilly mornings in July and August, and on 11 and 12 September and again on 27, hard frosts put an abrupt end to the already shortened growing season. Across eastern North America, summer temperatures averaged 3 to 5°C below today's normal temperatures. It has been just as cold or even colder in each of these months in other years, but never consecutively.

Over the years people have attributed the abnormal weather to an extraordinary increase in sunspot activity, changes in the phase of the moon, or earthquakes. Today, climatologists believe "the year without a summer" was a result of the eruption of the Tamboro volcano east of Java, which between April 7 and 12, 1815, ejected upwards of 150 million tonnes of dust, ash, and cinders into the atmosphere. The resulting global veil reflected incoming solar radiation back to space and indirectly chilled the earth.

northern regions, precipitation is much greater in the summer than in the winter.

Rain or snow days amount to about 150 or more across the province. Surprisingly, the number of wet days is greater in the centre and north of the province, where total

retreats slowly northward until mid-May. By mid-June, only the Ungava Peninsula remains covered. The average ice-free period for lakes ranges from 90 to 120 days in the north to 240 in the St. Lawrence Lowlands.

the month being foggy. Fog days number 60 a year in the southern Laurentians and 20 to 40 a year in the St. Lawrence River Valley, with a slight late summer and early fall peak in both regions. Settlements in the interior have a relatively low incidence of fog

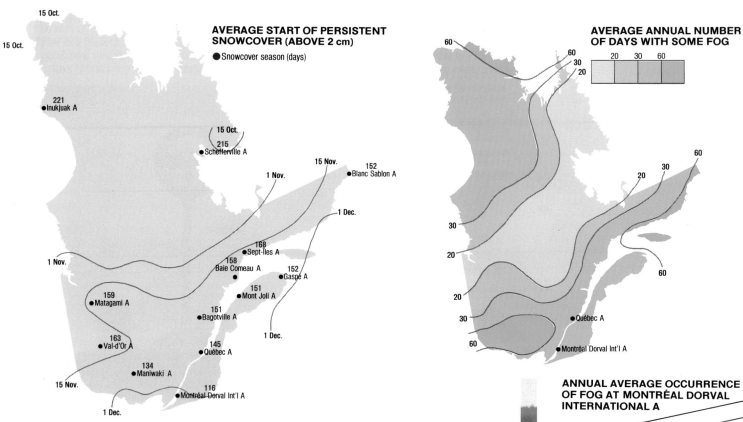

AVERAGE START OF PERSISTENT SNOWCOVER (ABOVE 2 cm)
● Snowcover season (days)

15 Oct.

15 Oct.

221
● Inukjuak A

15 Oct.

215
● Schefferville A

15 Nov.

1 Nov.

152
● Blanc Sablon A

1 Dec.

168
● Sept-Iles A

158
Baie Comeau A ●

152
● Gaspé A

1 Nov.

151
● Mont Joli A

159
● Matagami A

151
● Bagotville A

163
● Val-d'Or A

1 Dec.

145
● Québec A

134
● Maniwaki A

15 Nov.

116
● Montréal Dorval Int'l A

1 Dec.

AVERAGE ANNUAL NUMBER OF DAYS WITH SOME FOG

20 30 60

60
60
30
20

60

20

30

30

20

60

20

30

● Québec A

60

● Montréal Dorval Int'l A

precipitation is relatively light, than in the south. In the central interior around Schefferville, precipitation occurs on more than 200 days of the year.

Cold Cover

Snow comes early to many parts of Quebec. There is an even chance that the first snows will arrive in the extreme north by the end of September and over the whole central Shield country by the end of October.

The snow season gets started by mid-November on the elevated plateau to the north and west of the Gulf of St. Lawrence and over the northern mountains of the Gaspé Peninsula. Here is found the deepest snowcover in eastern Canada, with maximum accumulations each year of about 125 cm. Winter's first significant snowfall usually arrives at Montréal by November 10, and at Québec and the Eastern Townships by November 7.

In the spring, snow clears from the lowlands by mid-April. Owing to the great depth of snow to be melted, the snow-line

Misty Sunshine

Quebec is neither the sunniest nor the cloudiest place in Canada. The yearly total of bright sunshine ranges from about 1400 hours in the north to nearly 2000 hours in the south, or slightly less than 50% of the possible total sunshine. In July, the sunniest month, most of the province receives over 200 hours of bright sunshine. The sunniest area is northwest of Hull with 280 hours; the least sunny is along the Hudson Strait coast, where fog and cloud keep totals under 175 hours in spite of the 18-hour days. Seldom during the summer is a day completely overcast or cloud free. November has the least sunshine of any month, with most locations averaging less than 3 hours of sunshine a day and a few less than 1 hour a day.

Fog occurs on 60 to 80 days or more at the coastal settlements in the far north and along the north shore of the Gulf and the Gulf islands. On the Hudson Bay coast at Kuujjuarapik, fog can be expected on 45 days a year. The summer months are the foggiest here, with 15 to 25% of the hours in

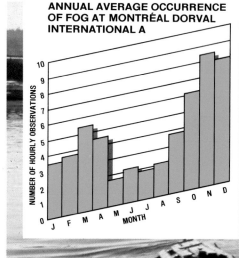

ANNUAL AVERAGE OCCURRENCE OF FOG AT MONTRÉAL DORVAL INTERNATIONAL A

NUMBER OF HOURLY OBSERVATIONS

10
9
8
7
6
5
4
3
2
1
0

J F M A M J J A S O N D
MONTH

— less than 20 days in an average year. At industrial settlements in the northern interior, such as Schefferville, ice fog becomes increasingly probable because of the presence of large amounts of water vapour produced by industries, homes, and cars. Ice fog occurs at temperatures below -30°C when the extremely cold temperatures cause water vapour to condense directly into tiny ice crystals instead of water droplets.

crossing North America, Quebec receives a full range of weather, including its share of quite violent storms. But foul weather of any kind seldom lasts for long because most storms hightail it out of the province within two days. Some thunderstorms produce heavy rainfall, and each year there are at least two or three reports of tornadoes and many more hailstorms. Hurricanes, with strong winds over 100 km/h and heavy flooding rains, have also been known to

affect Quebec. Winter storms — blizzards, freezing rain, and deep low pressure systems — are usually more destructive, however, and certainly create more wide-spread disruption than their summer counterparts.

Thunderstorms occur on about 20 days a year in southern Quebec, but on only one-half of that number in the east. In central parts of the province, the number is even lower. Annually less than one tornado per

Severe Weather
Being in the path of fast-moving storms

QUEBEC WEATHER SUPERLATIVES

	What	Where	When
Temperature			
highest	40.0°C	Témiscamingue	July 6, 1921
lowest	-54.4°C	Doucet	Feb. 5, 1923
Greatest ...			
precipitation	3125 mm	Mine Madeleine	1981
snowfall	1281 cm	Mine Madeleine	1980
number of fog days	113	Maniwaki	1968
wet period	252 days	Forêt Montmorency	1976
number of blowing snow days	120	Border Airport	1973-74
sunshine	2243 hours	Montréal Mirabel	1982
wind (maximum hourly)	201 km/h	Quaqtaq (Cape Hopes Advance)	Nov.18,1931
frost-free period	217 days	Montréal McGill	1963

AVERAGE ANNUAL NUMBER OF DAYS WITH SEVERE WEATHER

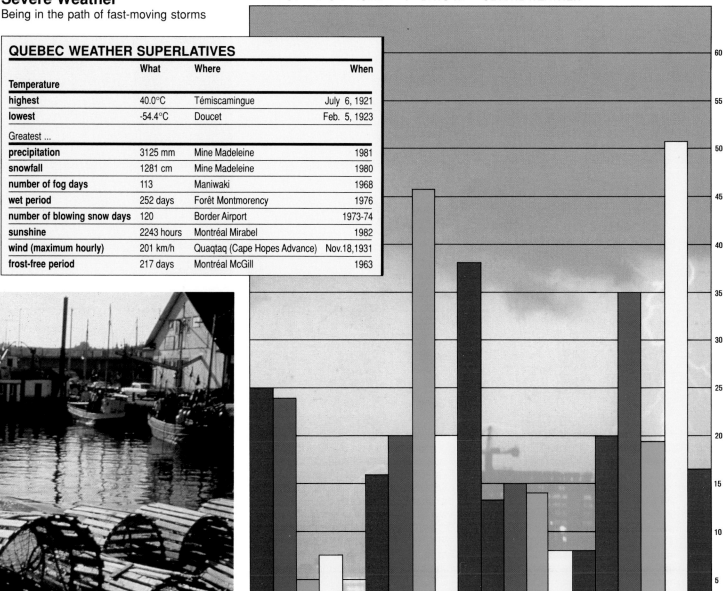

THUNDERSTORMS   BLOWING SNOW   FREEZING PRECIPITATION   FOG

● Montréal   ● Québec   ● Schefferville   Sept-Îles   ● Kuujjuaq (Fort Chimo)

10 000 km of land use can be expected in southern Quebec, compared with 1.5 to 2.0 in southern Ontario.

Hail is not usually a major problem. At Montréal Dorval, hail has been reported on only 34 days in 27 years, although in 1987 the airport had 25 thunderstorm days between May and August and hail occurred on at least 5 of them. Fewer than 1 in 10 thunderstorms has hail, and in some years there are no hailers. The worst hailstorm on record in the province occurred at Montréal

glaze storms cause enormous losses to trees, transmission towers, and cables. The most damaging and crippling ice storm ever experienced in Quebec hit Montréal in February 1961. Wires heavily loaded with 3 to 6 cm of ice snapped in winds of 90 km/h. Roads and airport runways were ice covered. Storm damage exceeded $7 million (equivalent to $34 million in 1990), but there was also uncounted personal discomfort and danger resulting from the loss of lighting and heating power for several days.

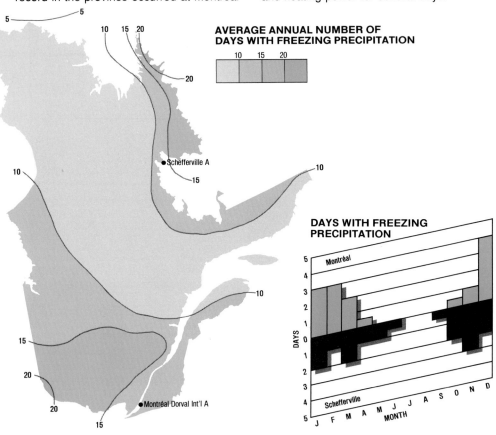

AVERAGE ANNUAL NUMBER OF DAYS WITH FREEZING PRECIPITATION

DAYS WITH FREEZING PRECIPITATION

on May 29, 1986. Hailstones of 10 to 12 mm diameter or more caused $50-75 million damage to trees, gardens, window glass, automobiles, and roofs. Exactly a year later, another multimillion dollar hailstorm pounded the city.

Some of the most damaging and costly glaze storms in Canada have occurred in Quebec. They are most frequent on the southern margin of the Laurentian Plateau, in the Gaspé, and in the high terrain of the Eastern Townships, where an average of 50 or more hours of glaze storms a year occurs, spread over 20 or more days. Occurrences decrease to less than 25 hours over central Quebec and less than 10 hours in Ungava. The region of greatest prevalence lies along the southern boundary of Quebec and Labrador.

When accompanied by strong winds,

Strong winter winds blow snow about in blizzard conditions on an average of 20 days each winter in the south, 30 days in central areas, and 60 days in the north of the province. In the cities, blizzards like the one that struck southwestern Quebec on February 19-20, 1972, tie up transportation for days and make travel extremely hazardous. The 1972 blizzard, which combined winds of 80 km/h and 37 cm of snow, was described as one of the most intense, widespread, damaging snowstorms ever to strike the province.

Storms that move northeastward along the Atlantic coast of the United States rarely pass directly over the province; however, their influence is undeniable. In late summer or early fall, remnants of tropical hurricanes that make it north to New England bring heavy rains and strong winds

to parts of southeastern Quebec. Hurricane Frederic, for example, brought 80 mm or more of rain to southern Quebec on September 13-14, 1979.

In winter, storms along the same track dump heavy snows and blow strong northeasterly winds over the southern and eastern parts of the province. These Atlantic coastal storms account for 70% of Montréal's big snowfalls.

Moving Air

Although topographic differences, sea breezes, and urban heat islands will cause variations in wind direction and speed from one observing point to another, there are common wind patterns in different regions of the province.

Exposed sites in the Gulf of St. Lawrence, such as Cap-aux-Meules on the

and in the Saguenay — Lac St. Jean region have stronger average wind speeds than any inland station in southern Quebec. Their extra strength may be due to funnelling in the deep river valleys. Such stations also show prevailing wind directions that follow the alignment of the valleys, southwest to northeast from Montréal to Québec, and west-northwest to east-southeast along the Saguenay.

Îles-de-la-Madeleine and the western tip of Anticosti Island, are some of the windiest places in Canada. The annual average hourly wind speed there is 31 km/h. Winds in the Gulf and around the estuary blow mainly from the southwest in summer and from the west-northwest in winter.

Farther north in the province, exposed sites along Hudson Strait and Hudson Bay have average yearly speeds of 21 km/h. In the interior, wind speeds are lighter, especially in less exposed areas such as forests, valleys, and hollows. Quaqtaq (Cape Hopes Advance) has the distinction of recording the highest 1-hour sustained wind speed in Canada — 201 km/h on November 18, 1931. On the same day a speed in excess of 160 km/h was recorded for 5 consecutive hours.

Stations along the St. Lawrence River

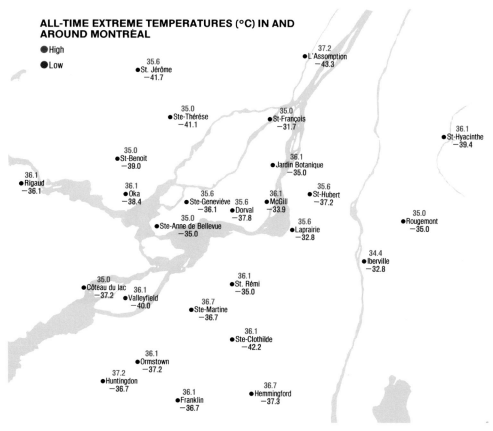

ALL-TIME EXTREME TEMPERATURES (°C) IN AND AROUND MONTRÉAL

● High
● Low

37.2
● L'Assomption
−43.3

35.6
● St. Jérôme
−41.7

35.0
● Ste-Thérèse
−41.1

35.0
● St-François
−31.7

36.1
● St-Hyacinthe
−39.4

35.0
● St-Benoit
−39.0

36.1
● Jardin Botanique
−35.0

36.1
● Rigaud
−36.1

36.1
● Oka
−38.4

35.6
● Ste-Geneviève
−36.1

35.6
● Dorval
−37.8

36.1
● McGill
−33.9

35.6
● St-Hubert
−37.2

35.0
● Rougemont
−35.0

35.0
● Ste-Anne de Bellevue
−35.0

35.6
● Laprairie
−32.8

34.4
● Iberville
−32.8

35.0
● Côteau du lac
−37.2

36.1
● Valleyfield
−40.0

36.1
● St. Rémi
−35.0

36.7
● Ste-Martine
−36.7

36.1
● Ormstown
−37.2

36.1
● Ste-Clothilde
−42.2

37.2
● Huntingdon
−36.7

36.1
● Franklin
−36.7

36.7
● Hemmingford
−37.3

PHIL DUDLEY

Moderate Consistent and Reliable

Even though Ontario's weather is variable, and it is that at all times of the year, its climate is consistent. To say it another way, there may be no area in Canada, east of the Rocky Mountains, that has a greater chance of getting the weather that's expected than Ontario.

Two in One

Ontario is often considered to have two distinct regions but many different climates.

Northern Ontario is Canadian Shield country. It occupies 90% of the province but contains only 10% of the population. Most of the land is forested, with intervals of swamp, muskeg, and small lakes. The infertility of the soil and the harshness of the climate have discouraged agriculture, but the economy of the area is sustained by mining, tourism, and lumbering. Northwest of Lake Superior the climate is similar to that in Manitoba — continental, with low winter and high summer precipitation and a range of 35 to 40°C between the average temperatures of the warmest and coldest months. There is some moderation of temperature in the Hudson Bay and James Bay area and around Lake Superior. As one travels south, temperature, rainfall, and snowfall increase but the range of temperature decreases. Across the north, frost-free periods range from 40 to 100 days, with local topography rather than regional climate being the key factor. Sunshine is ample, about the same as in the south, but somewhat less than on the Prairies.

Southern Ontario (the area south of Lake Nipissing) has a modified continental climate and experiences much less severe weather than the northern part of the province. In fact, its climate is one of the mildest and most benign of any region of Canada — a

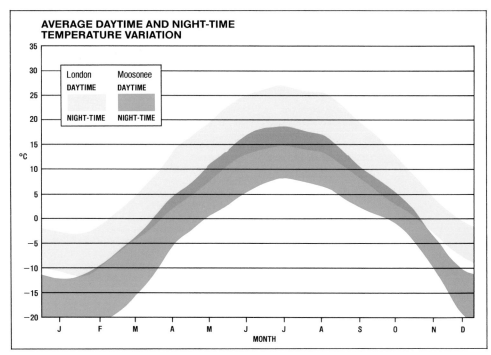

AVERAGE DAYTIME AND NIGHT-TIME TEMPERATURE VARIATION

factor that has contributed to the area's industrialization and inhabitation. Southern Ontario's climate owes much to the moderating presence of the Great Lakes and the region's southerly location. Point Pelee, its southernmost tip, is at the same latitude as northern California.

The unique features of the climate are four distinct seasons, a variety of precipitation types and sources, and a remarkably uniform distribution of precipitation from season to season. Warm summers prevail, with frequent uncomfortable periods of hot, humid, hazy weather. Warm, sunny days and crisp, cool nights make the fall season popular. Winters last 3 to 5 months, and

spring is normally the shortest season of the year, lasting from a few weeks to a couple of months.

The region lies across one of the major storm tracks of North America, and the passage of high and low pressure systems over the area produces wide variations in the day-to-day weather. Weather systems may be expected to traverse the region every two to five days throughout the year. Spells of weather — dry or wet, hot or cold — are seldom long.

Shaping the Climate

Besides the Great Lakes, several other important controls interact to shape both regional and local climates. Among the regional controls are latitude, air mass, storm movement, altitude, and topography. Local controls include urbanization, surface cover, and soil characteristics.

TOPOGRAPHY

The Ontario countryside south of the Shield consists of level terrain and gently rolling hills that permit the unobstructed movement of air

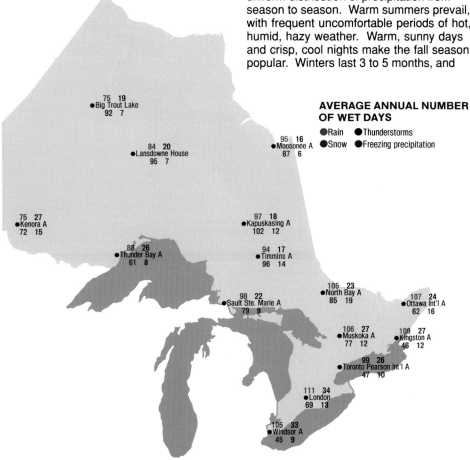

AVERAGE ANNUAL NUMBER OF WET DAYS

● Rain ● Thunderstorms
● Snow ● Freezing precipitation

THE GREAT LAKES

The Great Lakes are the world's largest single source of fresh water. Making up half of Ontario's southern and western borders, the giant lakes have a profound influence on the province's climate and especially on that of peninsular, southwestern Ontario.

Their influence is evident not just along their shores but is spread throughout their entire basin spread over 250,000 square kilometres. The lake effect also spreads downwind into adjacent regions. The most obvious effect is the moderation of temperature, but the lakes also intensify storms, increase precipitation and cloud cover in certain conditions (and decrease them in others), enhance wind speeds, and produce fog.

The lakes heat up and cool down far slower than the land does — changes in water surface temperatures lag one or two months behind changes in land temperatures. During the fall and winter, the surface water temperature is kept higher than the air temperature by the mixing of surface waters with warmer layers underneath. In spring, colder subsurface waters keep the surface water temperature well below the temperature of the overriding air. By late summer, the surface waters have warmed to their maximum. Lake Superior, the largest, deepest, and most northerly of the Great Lakes, reaches its maximum surface water temperature of 12°C in mid-September, about a month earlier than Lake Ontario reaches its peak of 20°C. In winter, Lake Superior develops an appreciable ice cover over half of its surface, whereas Lake Ontario may be only 10% ice-covered. Open lakes moderate winter weather more effectively than ice-covered bodies, which resemble land surfaces.

As a rule, summers near the lakes are slightly cooler and winters are warmer than in comparable inland locations. In delaying the arrival of the seasons, the lakes also prolong the growing season. This effect is crucial in spring in the fruit-growing areas of the region. The suppression of maximum temperatures near the shoreline delays blossoming until there is less chance of a damaging frost. Also, by extending the growing season in the fall and mitigating the harshness of winter, the lakes are a major factor in the success of southern Ontario's orchards and vineyards.

The lakes also affect precipitation. The most dramatic example is the abundance of snow downwind of the lakes. Stations in the so-called "snow-belt" receive from 250 to 400 cm of snow each winter, the product of moisture picked up by the prevailing winds in their passage over the lakes. The effect of the Great Lakes on warm season precipitation is less distinctive. There is evidence, however, that the lakes suppress thunderstorm development, especially in spring and early summer.

Although the lakes are usually a source of moisture for the air, they sometimes act in reverse. In the spring, for instance, when their surface waters are still quite cool, moisture is returned to the lakes through condensation from the overriding warmer air.

Other climatic effects created by the presence of the lakes include lake and land breezes near the shoreline; slightly more sunshine when the lake waters are cooler than the air over the land, more cloud when the lake water temperatures are warmer than that of the overlying air, and a higher frequency of fog.

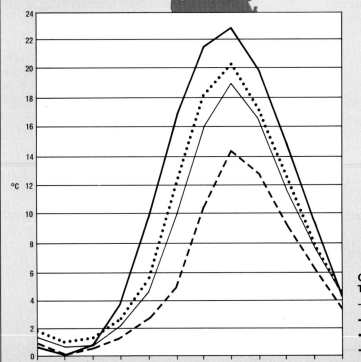

GREAT LAKES SURFACE WATER JULY TEMPERATURE (°C)

10 14 18

GREAT LAKES SURFACE WATER TEMPERATURE (°C)

——— LAKE HURON
——— LAKE ERIE
·········· LAKE ONTARIO
– – – – LAKE SUPERIOR

masses and storms during any part of the year. Differences in relief are not great, but topography does have a significant impact on local climate, especially in areas situated downwind of the Great Lakes. In the south, the two upland regions with large areas above 425 m in elevation — the Ontario Highlands, immediately east of Lake Huron, and the slopes of the Canadian Shield in Algonquin Park, east of Georgian Bay — are notably cooler, and, because of greater snowfall, wetter than the surrounding lowlands. In the north, other uplands are found around Lake Superior, with elevations of 600 m to the west of the lake and 450 m to the east.

One of the more prominent landforms in southern Ontario is the Niagara Escarpment. Beginning near Queenston, where its 100 m bluffs thrust above the level countryside, it continues along the south shore of Lake Ontario through Hamilton and northward to Georgian Bay, where the ski slopes of Collingwood reach elevations of 300 m or more. From there, it continues almost uninterrupted through the Bruce Peninsula and reappears on Manitoulin Island. Though not a massive natural barrier, it nevertheless has some significant local climatic effects. The most obvious of these is its capacity for wringing snow out of lake storms. Less dramatic but more welcome is its year-round power to dissipate clouds and precipitation, lessen winds, and warm the air masses that

AIR MASSES INFLUENCING CENTRAL ONTARIO

SUMMER

Maritime Arctic cool and dry (about 35%)

Maritime Polar warm and dry (about 45%)

Maritime Atlantic warm and humid (about 5%)

Maritime Tropical Hot and humid (about 15%)

WINTER

Continental Arctic very cold and dry (about 30%)

Maritime Arctic cold and moist (about 3%)

Modified Continental Arctic cold and moist (about 50%)

Maritime Arctic moist and mild (about 15%)

Maritime Polar moist and mild (about 2%)

sink on the downwind side of the escarpment to the east.

The physical feature that has the most profound affect on Ontario's climate, however, is probably the Appalachian mountain system, which lies outside the province along the eastern rim of the Great Lakes Basin. These mountains direct storms

moving along the Atlantic coast away from Ontario. However, on occasion, tropical storms push over the Appalachians and re-develop in Ontario, bringing heavy precipitation.

LOCATION
The southernmost tip of Ontario lies just south of the 42nd parallel of latitude. Its northern border lies above the 55th. Between them is some 1600 km, enough to make more than a one-hour difference in the length of day at the two extremities on the shortest and longest days of the year.

Located near the centre of North America, Ontario could be expected to have dramatic changes in temperature from winter to summer or from day to night, much like those that mark the continental climate of the Prairies. Instead it is a hybrid of continental and maritime regimes, with a less extreme temperature range than the Prairies and less precipitation than the Maritimes.

WEATHER SYSTEMS
Ontario's climate features a steady progression of settled and stormy weather on an almost day-to-day basis throughout the year. Good or bad periods usually do not last for more than a few days. The settled weather is associated with high pressure systems that bring extensive masses of air from such diverse sources as the tropics, the Arctic, and the Pacific and Atlantic oceans. Shallow, cold, and dry Arctic air predominates in the north during the winter. But southern Ontario also receives its share of clear, cold and sunny days under Arctic air, which,

PHIL DUDLEY

according to estimates, is present 22% of the time in winter. Atlantic air is associated with cloudy skies, and Pacific air generally brings a varied package of cool, moist, cloudy, and dull weather or, at times, sunny and mild weather. Tropical air enters the south regularly in summer but rarely in winter, at least at ground level. This air is characterized by oppressively high temperature and humidity, high pollution concentrations, and afternoon thunderstorms. Tropical air occurs about 30% of the time over the lower Great Lakes in summer, but less than 10% of the time north of Lake Superior.

Along the fronts that separate these contrasting and competing air masses, storms develop. They favour a northeasterly path through southern Ontario on their way out to the Gulf of St. Lawrence and the North Atlantic. Storms are more vigorous in winter, bringing strong winds and a pot-pourri of precipitation types — rain, freezing rain, ice pellets, and snow, or a mixture of all of them. While the storms depart the province, they leave behind strong winds, low temperatures, and blowing snow. In spring and summer, cold fronts may trigger violent thunderstorms and, infrequently, tornadoes. Thunderstorms supply much of the summer rainfall.

The Hot and Cold of It

Temperatures have exceeded 38°C at nearly all Ontario localities. The province's highest temperature is 42.2°C, recorded at Atikokan and Fort Francis in the second week of July 1936. At the same time, Toronto's temperature was 40.6°C. These temperature records were set during one of the severe heat waves that periodically afflict Ontario in the summer. The 1936 heat wave, which gripped most of western and central Canada from Alberta to the Ontario — Quebec border, was one of the most intense in Canadian history. Approximately 600 deaths in Ontario, almost half of them in Toronto, were attributed to the high temperatures, including several from heat stroke and some from drownings. In addition, crops were blackened, fruit was baked on trees, and tinder-dry forests were set ablaze.

The lowest temperature ever recorded in Ontario was -58.3°C on January 23, 1935, at Iroquois Falls. The range of average temperature, from the coldest to warmest month, is about 35°C in the north — comparable to the temperature range experienced on the Prairies. In the south, the contrast between summer and winter temperatures is between 25 and 30°C.

Generally, temperature severity increases as one moves east and north in Ontario. Winter temperatures are highest along the

ONTARIO WEATHER SUPERLATIVES

	What	Where	When
Temperature			
highest	42.2°C	Atikokan	July 11/12, 1936
		Fort Francis	July 13, 1936
lowest	-58.3°C	Iroquois Falls	Jan. 23, 1935
Greatest			
precipitation	1620 mm	Stratford	1884
snowfall	766 cm	Steep Hill Falls	1938-39
number of fog days	99	Mount Forest	1978
wet period	232 days	Burks Falls	1980
number of blowing snowdays	52	Winisk Airport	1962-63
sunshine	2454 hours	Thunder Bay	1976
wind (maximum hourly)	114 km/h	Caribou Island	Nov. 22, 1946
frost-free period	228 days	Pelee Island	1948

Great Lakes. Even in northern Ontario, the warming effect of Lake Superior keeps average temperatures 3 to 5°C above those at stations inland. North of Superior, winters are extremely cold, averaging -20°C in January from Lake of the Woods to south of James Bay. Hudson Bay and James Bay freeze over, neutralizing any modifying influence that open water would have on the cold Arctic air. Consequently, across the far north average temperatures of -25°C occur, comparable with those found in the northern Prairies, and extreme temperatures can reach as low as -50°C.

Along the north shore of Lake Erie and in the Niagara Peninsula, extreme winter low temperatures seldom fall below -25°C, which is an important factor for fruit growing in the area. Along the St. Lawrence River and in Eastern Ontario, winter extremes dip to -35°C and mid-winter averages are around -10°C.

The temperature differences across the province are much less pronounced in summer than in winter. Further, the effect of the lakes is less marked and less extensive in July when the average daily air temperature is very similar to the surface water temperature of the Great Lakes. In the extreme southwest, the Windsor – Leamington area is the warmest, with afternoon temperatures averaging above 27°C. Windsor often vies with interior British Columbia for the distinction of having the nation's warmest summer month. Towards James Bay, summer afternoon temperatures exceed 20°C, but the season is brief. In the Georgian Bay, Muskoka, and Haliburton cottage country, July temperatures average 18 or 19°C but are a shade warmer across the Kawarthas and the northern shore of Lake Ontario. Around Lake Superior, average daily temperatures consistently stay near 16°C, but a short distance away they are near 20°C, not unlike those in the south. Also it is not unusual for the north, like the south, to have periods of hot weather with occasional occurrences of maximum temperatures in excess of 32°C. However, hot spells across the province are generally short-lived, lasting days not weeks. A notable exception occurred in the summer of 1988, when 31 days had temperatures exceeding 30°C; the highest number of hot days in almost 40 years.

Similarly, in the south, winter cold is seldom locked in beyond ten days. The regular west to east procession of weather systems during the year ensures that prolonged periods of hot or cold are infrequent.

Spring and fall, the seasons of transition, are often the most pleasant seasons of the year, offering daytime warmth and night-time coolness. They are short-lived, however, and never last as long as three months. Spring 1985 was especially fleeting — it lasted just one week. In mid-April, the last winter storm brought snow and gale-force winds to the south. A week later, the thermometer registered 30.3°C at Toronto, a new monthly record and, as it turned out, the warmest temperature all year.

In the fall, freezing temperatures can be expected in the south sometime in late September or October, heralding the arrival of Indian summer — that welcome spell of sunny, hazy, warm days and frosty nights that occurs each fall in eastern North

America. Near the lakes, the rush of early fall frost is lessened and the growing season is lengthened.

In discussing temperatures, it is important to consider the urban effect that keeps the core of Canadian cities in midwinter 2 to 3°C warmer on the average than the surrounding rural areas. On occasion, these differences exceed 10°C.

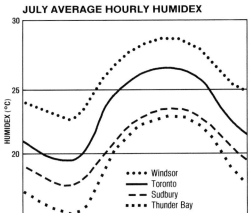

JULY AVERAGE HOURLY HUMIDEX

- Windsor
— Toronto
-- Sudbury
··· Thunder Bay

Frost-Free

On average, the last killing frost in southern Ontario occurs about the first week in May, and the first fall frost occurs during the first half of October. This leaves a frost-free season of 160 days. Point Pelee's 180 days is longer than that of any other place in Canada, except along the Pacific coast, and is similar to values found across northern Missouri and southern Ohio. Around the lakes, frost-free periods are one to two months longer in duration than inland. The season shortens to 140 days along the St. Lawrence and 100 days in Algonquin Park, Haliburton, and the northern shores of Lake Superior and Lake Huron. Elsewhere, frosts can be expected until late June in the far north and until the beginning of June around Sault Ste Marie. They reappear in late August in the north and during the last week of September near the Sault, making for a frost-free season of only 50 to

AVERAGE FROST-FREE PERIOD (DAYS)

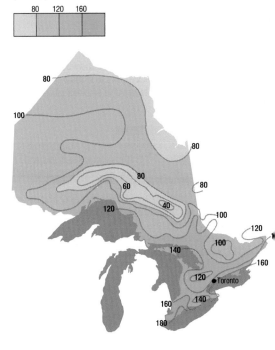

120 days.

Frost dates vary considerably over short distances because of topography, lake effects, vegetation, soil type, and different types of land use.

Southern Discomfort

Across most of Ontario there may be one or two periods of hot and humid weather each year when most people begin to suffer. A humidex reading of 30°C causes some people discomfort, while 40°C makes most people feel uncomfortable.

AVERAGE ANNUAL SNOWFALL (cm)

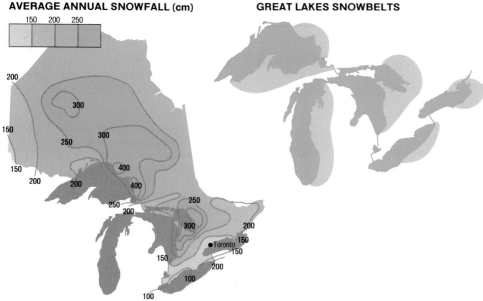

On average, southwestern Ontario experiences the highest frequency of hot, humid weather in Canada, usually during the latter two weeks of July. The highest values occur at Windsor where, at the peak humidex hour of 3 p.m., the mean humidex during the last week of July is 32°C. Extreme values above 48°C have occurred at several locations in southwestern Ontario. Generally, humidex decreases with increasing latitude. Lower values are also found at higher elevations, near the shores of lakes, and in forests.

Reliable Wetness

There are few surprises with precipitation across Ontario. Precipitation extremes are minimized, amounts are evened out, and wet or dry spells are comparatively rare. In one word, rain and snow in Ontario are about as "reliable" as one could find in a temperate continental climate.

Precipitation over southern Ontario is fairly uniformly distributed throughout the year. It is this seasonal uniformity that chiefly distinguishes this region's climate from others in Canada. The contribution

from the more vigorous and frequent storms of winter matches that from less developed low pressure systems and thunderstorms in the warmer, moister air of summer. If anything, more rain falls during the growing season than at any other time of year. February is clearly the driest month in many areas, but in some instances this is only because it has the fewest number of days.

Northern Ontario, with its more continental climate, receives its maximum precipitation in summer. Quorn, 90 km north-northeast of Atikokan, has three and a half

GREAT LAKES SNOWBELTS

times more rain in summer than it has rain and snow combined in winter.

The amount of precipitation is affected by location relative to the Great Lakes, prevailing winds, elevation, and slope of the land. The greatest amounts occur on highlands downwind of the lakes. The south is about one third wetter than the north, with totals ranging from 500 mm in the Patricia area of northwestern Ontario to 1000 mm and more on the western flanks of the highlands of Lake Superior, Lake Huron, and the Georgian Bay. Another area of heavy precipitation is along the St. Lawrence River.

The driest parts of the south are the fruit-rich areas of Essex County, in the extreme southwest, and the Niagara Peninsula. Similarly low levels of precipitation occur also in the rain-shadows of the highlands — the eastern half of Algonquin Park towards Pembroke, for example, and the Simcoe-Toronto corridor east of the Dundalk Highlands.

For the record, the wettest station on average (1951-1980) is Chatsworth with 1077 mm, and the driest is Big Trout Lake with 581mm.

Precipitation reliability also holds true from year to year. Periods of excessively dry or wet weather are not frequent, although periods of drought are more likely to occur in the late summer, in time for the final harvest, than at any other time. Dry periods of one month or longer are rare, but those of one week are not, occurring on average at least once a month during the growing season. Recent droughts have occurred in 1973, 1978, 1983, 1988, and 1989.

There is little variation around the province in the number of wet days each year. An average for the province is 125 wet days, with fewer in the extreme southwest, and about 150 in areas to the lee of the Great Lakes.

Ontario Snowbelts

Ontario's snowbelts are legendary. Skiers swarm to these regions every weekend in winter and early spring to enjoy the snow. On the upland slopes facing Lake Huron, Georgian Bay, and Lake Superior, huge snowfalls in the 300 to 400 cm range occur each winter, with much of it coming from the prevailing winds blowing off the open lakes. The cold air is warmed and moistened while it travels over the lakes. It is then forced to ascend the highlands, triggering heavy snows. Also adding to the snow totals is snow squeezed out of winter storms, while they rise over the higher terrain of the province. This orographic effect is estimated to contribute about 17 cm of snow for each 30 m rise of elevation. The highlands south of Owen Sound, around Parry Sound, and northeast of Sault Ste Marie are snow-favoured as are the areas along the St. Lawrence River and around London.

On occasion, Ontario snowstorms can pack a heavy punch. One of the worst in recent memory occurred in London between 7 and 9 of December, 1977, when 101 cm of snow fell with winds howling to 100 km/h. The drifting snow necessitated a state of emergency and the mobilization of the armed forces to restore essential services.

Areas to the lee of the lakes but on the downslope side of the higher ground receive less than half the annual snow totals of the upslope snowbelt areas. Toronto, Hamilton, and places in view of the east side of the Niagara Escarpment, and the eastern Algonquin Park area are snow-shadow regions. Smaller amounts also occur in the southwest in Essex County (100 cm), over the Niagara Peninsula (115 cm), and in the northwest along the Ontario — Manitoba border (150 cm).

Snowfall accounts for about 33% of the year's total precipitation in the snowbelt regions. In the snow-sparse area around Windsor and Chatham, the amount falls to 12%.

PETER ELMS

A general characteristic of Ontario snowfall patterns is that they are highly variable. This is especially true in the southern section of the province, where a single storm may bring as much or more snow to a locality in one or two days than it would receive in an entire winter some other year.

The number of snow days (i.e., days with measurable snowfall) exceeds 75 a year on average in the snowbelts. There are between 75 and 100 snow days each year from Lake Superior to the shore of Hudson Bay. Over 40 days of snow occur annually on the north shore of Lake Ontario, and there are 50 snow days in the Ottawa Valley. These values contrast with fewer than 20 days a year at Point Pelee in the extreme southwest.

White Cover

It does not necessarily follow that the place with the highest snowfall will have either the deepest or the most prolonged snowcover. The deepest snowcover, over 100 cm,

TORONTO TELEGRAM - YORK UNIVERSITY

METEOROLOGICAL MOMENT

Undoubtedly more words have been written about Hurricane Hazel than any other single weather event in Canadian history. After carving a path of destruction through the Caribbean, Hazel struck the American coast near Myrtle Beach on the morning of October 15, 1954. Over the next 12 hours she raced from North Carolina to southern Ontario. Weakened by the overland journey, and with peak winds less than 115 km/h, Hazel was technically no longer a hurricane, but after merging with a cold front west of Toronto she regained some of her lost fury.

On the evening of the 15th, Hazel struck an unprepared and unattentive Toronto. Eighteen hours later, she departed for Hudson Bay, but not before claiming 80 lives and causing over $ 130 million (in 1990 dollars) of storm-and-flood-related damage. In its wake, the storm left a nightmare of destruction — lost streets, washed-out bridges, and untold personal tragedy. The greatest loss of life and property damage occurred in the Humber River Valley and in the Holland Marsh just south of Lake Simcoe. Unprecedented downpours of up to 178 mm easily saturated well-watered soils, producing flash flooding on the rivers and creeks.

Apart from the two-day rainstorm that flooded parts of Essex County in July 1989, no other storm before or since has challenged Hazel's status as the heaviest rainstorm ever recorded in southern Ontario. And none has produced such tragic and dire memories.

occurs around Lake Nipigon and east of Lake Superior. The end of January is normally the time when there is the most snow on the ground in central, southern, and eastern Ontario, but it is the first of March across northwestern Ontario.

A persistent snowcover usually forms gradually from north to south as follows: end of October in northwestern Ontario (Winisk and Severn rivers); November 15 around Lake Superior and east to Sudbury; first week of December from London east to Kingston; and around Christmas time in extreme southwestern Ontario, where it is at best intermittent throughout the winter.

Snowcover is normally gone by April 1 in the south, mid-April in the snowbelt, and the end of April across central Ontario. In the north, snow disappears by mid-May, making for a snowcover season of over half a year. What northern Ontario lacks in amount of snowfall, it makes up in the length of the snow year. A good portion of the northern Ontario total of 150 to 250 cm stays throughout much of the year.

Stormy Weather

Ontario lies in a zone of active weather movement. Winter storms are usually more vigorous and more frequent than those of summer. Weather variety is the norm, and because systems move so rapidly, spells — stormy or quiescent — never last very long.

About every other year in late summer or early fall, the southern and central portions of the province are affected by the remnants of tropical hurricanes while they move north-eastward up the Atlantic coast or, on rare occasions, move directly over the province. About once in six years hurricanes in Ontario make "front-page" news because of the damage caused by strong winds and/or excessive rainfall. Hurricane Hazel is the best remembered of these tropical cyclones.

Intense winter storms are frequently accompanied by numbing cold, ice or glaze, and heavy snow. One or two bad blizzards strike the province each winter, snarling traffic and forcing numerous delays and cancellations. Heavy icing, especially when accompanied by strong winds, creates extremely hazardous conditions and causes considerable damage to trees and power lines. Ottawa's 14 hours of freezing rain on Christmas Day, 1986, for example, inflicted extensive property damage and forced thousands to alter holiday plans. Occurrences of freezing precipitation vary from 10

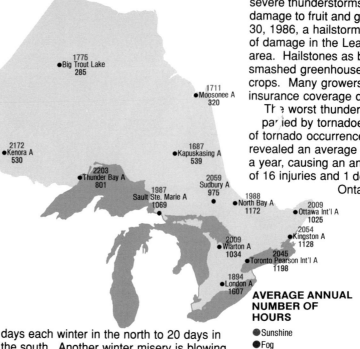

On occasion, hail accompanies the more severe thunderstorms and can cause heavy damage to fruit and garden crops. On May 30, 1986, a hailstorm inflicted $30-40 million of damage in the Leamington — Windsor area. Hailstones as big as golf balls smashed greenhouses and flattened early crops. Many growers subsequently found insurance coverage difficult to obtain.

The worst thunderstorms may be accompanied by tornadoes. A study of 30 years of tornado occurrences across Ontario revealed an average of 11 twisters a year, causing an annual average of 16 injuries and 1 death. Southwestern Ontario experiences the highest frequency and greatest risk of tornado damage of all areas in Canada — an average of 1 or 2 tornadoes per 10 000 km annually. Even more disturbing is the fact that a disastrous tornado can be expected somewhere in the south once every five years on average. Tornadoes occur less frequently, however, near the shores of the Great Lakes, in the north, and along the lower St. Lawrence River.

Map data (Average Annual Number of Hours — Sunshine):

- Big Trout Lake 1775 / 285
- Moosonee A 1711 / 320
- Kenora A 2172 / 530
- Kapuskasing A 1687 / 539
- Thunder Bay A 2203 / 801
- Sudbury A 2059 / 975
- Sault Ste. Marie A 1987 / 1069
- North Bay A 1988 / 1172
- Ottawa Int'l A 2009 / 1025
- Kingston A 2054 / 1128
- Wiarton A 2009 / 1034
- Toronto Pearson Int'l A 2045 / 1198
- London A 1894 / 1607

AVERAGE ANNUAL NUMBER OF HOURS
- ● Sunshine
- ● Fog

days each winter in the north to 20 days in the south. Another winter misery is blowing snow, which occurs 10 to 20 days each year. Snowbelt areas to the lee of the Great Lakes experience the worst and most frequent blowing snow conditions.

Thunderstorms can be expected to occur on 25 days each year south of Lake Superior and on as many as 35 days in the London — Windsor region, which offers more of nature's sound and light shows than any other area in Canada. The frequency drops off rapidly to the north, and along the shore of Hudson Bay it is 5 days or less.

Sun Parlour

As a region, southern Ontario receives less sunshine annually than other areas at comparable latitudes. This is because of its location in the humid and industrial region of North America. The presence of atmospheric pollution from its cities and its

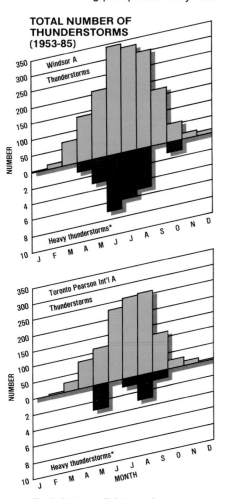

TOTAL NUMBER OF THUNDERSTORMS (1953-85)

Windsor A — Thunderstorms / Heavy thunderstorms*
(NUMBER vs. months J F M A M J J A S O N D)

Toronto Pearson Int'l A — Thunderstorms / Heavy thunderstorms*
(NUMBER vs. MONTH J F M A M J J A S O N D)

*Heavy thunderstorms occur with sharp pronounced and continuous thunder and lightning . . . heavy rain, peak winds above 63 km/h and a rapid drop in temperature usually accompany the storm. Hail or snow may occur.

proximity to the cloud-forming Great Lakes contribute to decreased sunshine amounts. Most of the province receives 1800 to 2000 hours of sunshine annually, or 45% of the possible total. Only in the far north, near James Bay, is the total less than 1800 hours.

July is the sunniest month, having an average of 9 hours of sunshine a day. Ottawa averages only one completely overcast day in July. However, there are 5 sunless days in October, 9 in January, and 5 in April. Fall is generally the cloudiest season. Even Canada's so called "sun parlour", Essex County, receives less than 3 hours of sunshine a day in winter.

Fogged In

Fog in Ontario tends to be more frequent in the south-central area of the province. Annual fog days number 76 at Mount Forest, 69 at Sudbury, and 63 at North Bay — not nearly as foggy as Canada's East Coast but considerably foggier than the Prairies and about as foggy as British Columbia's coast. The occurrence of dense fog decreases to 15 days or less annually in the northwest around Red Lake and Pickle Lake. In the south, 35 days a year is a typical average.

Across the north, summer fogs prevail. In the south, the foggiest time is fall and winter when night-time cooling of the land produces early morning fog during light winds, and when steam fog forms over the lakes under cold, dry air. In the spring, fog sometimes forms directly over the lakes when passing streams of warm air are cooled by the cold lake waters. Later, when surface waters have become warmer, lake fogs may still occur when cold waters from lower depths rise to the surface and come in contact the warmer air above.

Wind

Winds are at their maximum in spring, when vigorous disturbances pass through the province. The lightest winds occur in the summer. In winter, average wind speeds increase by about 4 km/h. Lake and land breezes are common during the summer around the shorelines of the Great Lakes, with the winds usually blowing off the water during the day and off the land at night.

The prevailing winds are generally from the west, with southwest winds dominating in summer and west and northwest winds blowing most frequently in winter. Easterly winds are least frequent and the lightest. A notable exception is found at Sault Ste Marie, where passing winds are channelled between a ridge of hills to the northwest of the airport and the St. Mary's River to the south, causing them to pick up speed and blow from the east.

In valleys, wind speeds are noticeably lower than they are on highlands and along the shores of the Great Lakes and other, smaller lakes. Most sites in Ontario have recorded winds exceeding 100 km/h. Such heavy winds are usually associated with thunderstorms, squall lines, frontal passages, and strong low pressure systems.

Over the Great Lakes, November is usually the stormiest and therefore the windiest month, yet the records for extreme one-minute speeds on the various lakes, have all been set in June, July, and August. However, ships are not on the lakes in the winter months when stronger winds might be observed.

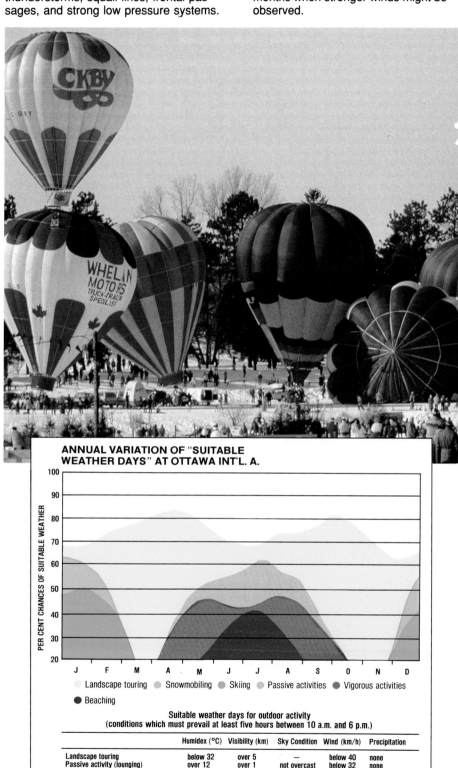

CHRIS STEWART

ANNUAL VARIATION OF "SUITABLE WEATHER DAYS" AT OTTAWA INT'L. A.

PER CENT CHANCES OF SUITABLE WEATHER

○ Landscape touring ● Snowmobiling ● Skiing ● Passive activities ● Vigorous activities
● Beaching

Suitable weather days for outdoor activity
(conditions which must prevail at least five hours between 10 a.m. and 6 p.m.)

	Humidex (°C)	Visibility (km)	Sky Condition	Wind (km/h)	Precipitation
Landscape touring	below 32	over 5	—	below 40	none
Passive activity (lounging)	over 12	over 1	not overcast	below 32	none
Vigorous activity (games, etc.)	12-32	over 3	not overcast	below 32	none
Beaching	over 18	over 1	not overcast	below 25	none
Skiing*	over −14	over ½	—	below 25	little or none
Snowmobiling*	over −21	over ½	—	below 25	little or none

*two centimetres or more of snow cover is needed.

True Canadian Climate

METEOROLOGICAL MOMENT

The Red River flood of May 1950 was the greatest flood disaster in Canadian history. Higher crests have occurred before along the north-flowing Red River, but the duration of the 1950 flood was much greater and the losses much heavier than on any other occasion. Losses were enormous—5000 homes and buildings were damaged, 100 000 people were evacuated and damage costs exceeded $ 550 million (in 1990 dollars). More than 1600 square kilometres of rich farm land and nearly a quarter of Winnipeg were under water. The centre of Winnipeg was spared because of an airlift of three million sandbags put into place by its tireless citizenry. For two weeks the river gauges recorded 4 m above the normal high water level, but by May 21 the river began to recede.

What caused the flood? Several factors were responsible: soggy soils from heavy fall rains before freeze-up, heavy winter snows, a late spring break up on April 16, and heavy rainfall before break up.

To prevent future disasters of this magnitude, a 47-kilometre floodway was later built to divert flood waters around the city.

Manitoba lies in the heartland of North America. Winnipeg, its capital, is more than 2000 km from the Pacific Ocean and more than 3000 from the Atlantic. Even the shorter of these distances is made effectively greater, in climate terms, by the western Cordillera. As a result of its passage over these mountains, Pacific air is substantially changed by the time it reaches Manitoba.

Manitoba's climate is purely continental. Its chief distinguishing marks are

- enormous temperature differences from summer to winter (the greatest in Canada)
- long, hard winters and warm, short summers

- low, but notoriously variable, annual precipitation totals
- dry winters and summers, with slightly more precipitation in summer

Climate Controls

Even though one sixth of Manitoba's surface is covered by lakes and rivers, water does not exert a major influence on the province's climate. Some lake effects do show up, however, in the region between Lake Winnipeg to the east and lakes Winnipegosis and Manitoba to the west. They result in slightly cooler summer temperatures and a temperature range that is generally smaller than that of other places at the same latitude. As for Hudson Bay, its modifying effect does not extend much

beyond its own shore on the Manitoba side. Churchill is about 3°C cooler in summer than places 10 km away from the coast but, because of extensive ice cover, receives little benefit in the winter from the heat stored in the waters of the Bay.

Since Manitoba is mostly a vast, gently sloping plain, terrain is not a significant climate control either. The only relief feature with some effect on climate is a series of minor upland areas between 700 and 800 m high along the Manitoba Escarpment near its western boundary with Saskatchewan. In fact, the absence of a substantial west-to-east ridge to the north of the Prairies leaves Manitoba open to frequent incursions of Arctic weather.

Water and relief features outside the province actually have a much greater influence on the province's climate. The mountains of the western Cordillera, for example, remove most of the moisture from the mild Pacific air passing over them and heat the westerly winds descending the eastern slopes of the Rockies. Nevertheless, at all times of the year much of Manitoba's moisture still comes from the Pacific, as does the occasional mild spell in winter. In summer the Gulf of Mexico also plays a part. It is a significant source of moisture for Manitoba thunderstorms.

Good Agricultural Weather

Agriculture in Manitoba employs more people and adds more value to the economy than any other resource industry. It produces more than a billion dollars worth of grains, specialized crops, and livestock each year. Climate and soil have much to do with what crops are grown and where. For example, wheat is common throughout southern Manitoba, but prime malting barley prefers the cooler summer temperatures of the southwest parkland with its mixture of grassland and woodland cover, and sugar beets, sunflowers, corn, and canning vegetables favour the greater warmth of the Red River Valley.

Summer temperatures generally range from the low teens at night to the mid-20s in the afternoon — the greatest daily range of temperature anywhere in Canada. Average midday temperatures are similar to those in southern Ontario, but average night-time temperatures are as much as 5°C lower in southern Manitoba. Since humidities are lower as well, Manitobans have a much better chance of a good night's sleep than their counterparts in southern Ontario.

Fortunately for agriculture, 60% of the annual precipitation occurs in the growing season from May to August, that is, at the time of year when it is of most benefit to the crops. In favourable years, rains will

JIM OLEKSUK

let up in late August and September during the harvest. June is the wettest month, owing to a greater frequency of weather disturbances that are characteristic of spring. Between June and July, a marked northward shift in the storm track occurs, bringing a reduction of precipitation.

Of the three Prairie provinces, Manitoba is the most favoured for precipitation, with totals ranging from 400 mm in the north to about 600 mm in the southeast. Manitoba's more abundant and generally more reliable precipitation results from its greater exposure to southerly flows of warm moist air from the central United States and the Gulf of Mexico.

The growing season starts early, during the second half of April, and usually ends early in the first half of October. The number of growing degree-days above 5°C varies from a maximum of 1900

south of Winnipeg to fewer than 550 near Hudson Bay. Most agricultural areas in Manitoba have between 1700 and 1800 degree-days, which is 100 to 300 higher than totals in Alberta and Saskatchewan.

Agriculture is frequently at risk because of early fall frosts. September 20 is the average date for the first frost in southern Manitoba. North of Lake Winnipeg the first frost usually occurs by September 10. The longest frost-free season, 130 days, occurs between Morden and Portage la Prairie, within the area influenced by Lake Manitoba. Thompson is particularly frosty, with only 59 frost-free days in an average year.

The occasional stream of warm, humid air from the Gulf of Mexico keeps humidities above the dry moisture levels of southern Alberta but well below the often oppressively high readings of southern Ontario. High evapotranspiration reduces the available moisture, especially in the southwest where some irrigation is practised.

Average water deficiencies for crops are about 100 to 150 mm, and although this is worrisome to Manitoba farmers, it is only one third to one half of what farmers face in the dry belt of southeastern Alberta and southwestern Saskatchewan. Droughts are neither as frequent nor severe as they are in the neighbouring province of Saskatchewan.

Sunshine totals soar in Manitoba, compared with parts of eastern Canada, but they are less than to the west. Annual totals exceed 2200 hours in the south, but around Churchill the figure drops to 1800 hours, in part because of the influence of Hudson Bay. July is usually the sunniest month, with average daily totals exceeding 10 hours at several stations.

Six Months of Winter
Cold outbreaks of Arctic air, which periodically affect North America during the winter

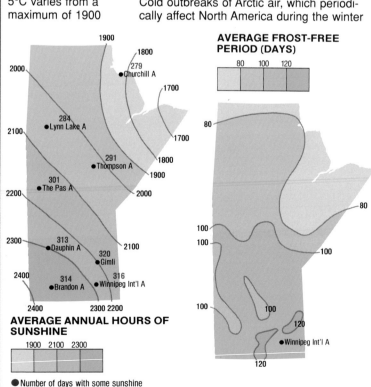

AVERAGE ANNUAL HOURS OF SUNSHINE

1900   2100   2300

● Number of days with some sunshine

AVERAGE FROST-FREE PERIOD (DAYS)

80   100   120

months, move unobstructed across Manitoba. The cold is not altered to any appreciable extent by the time it reaches Winnipeg. On occasion, incursions of mild Pacific air push across as far east as and sometimes beyond southern Manitoba. The result is a great variation of climate from month to month during a single winter or from year to year, depending on the character and origin of air masses passing over the province. Even daily temperatures may vary substantially. A high of -25°C one day may be replaced by a high of 5°C the next.

Winter's sweep of cold Arctic air can be awesome. Manitoba's coldest temperature, -52.8°C, was recorded at Norway House on January 9, 1899. January daytime/nighttime temperatures average about -14/-25°C in the southern grain area, but in the north averages of -24/-32°C are typical. Winter temperatures are not only severe, but the cold is of extremely long duration. On March 16, 1979, the longest spell of below-zero temperatures in southern Manitoba ended after 126 days. Minimum temp-

eratures reach freezing or below on 200 days of the year in the south and 250 days in the north. "Raw" days, with minimum temperatures below -20°C, occur on an average of 65 days a year in Winnipeg and 115 at Churchill and farther north.

There are mild winters also, when warmer weather persists well into December. Winter 1986-87 was the warmest on record, almost 7°C above normal. These years are rare, however. More often, snow and cold arrive in the south by mid-November and last until at least the middle of April. In the north, winter's start is a month earlier and its close is a month later. January is the coldest month throughout the province, with the period between January 15 and 25 the deepest part of winter. But by February 9 the temperature begins its long, slow rise toward springtime values.

Never A Green Christmas
Snowfall is not heavy — it just seems that way. Once the snow arrives it normally stays. Snow has fallen in Manitoba in every month but July. Winnipeg's seasonal

snowfall total of 125 cm is about 25 cm short of the provincial average, although in some years the capital has received as much as 253 cm (1955-56) and as little as 31 cm (1877-78).

Wind Chilling The Air
Winds are usually not as brisk in winter as they are in the spring and fall across Manitoba, and this is fortunate because the wind chill discomfort caused by the combination of low temperature and wind is reduced. Wind chill is a simple measure of the chilling effect experienced by the human body when winds are combined with cold temperatures. The loss of body heat in cold air increases with a rise in wind speed.

Ten per cent of the time in January, wind chill values exceed 2000 W/m² (equivalent to a still-air temperature of -42°C) in southern Manitoba and 2200 (equivalent to -48°C) in northern Manitoba around Churchill. Extreme wind chills, ranging from 2300 to 2600 W/m² (-53 to -64°C), occur 1% of the time and present a life-threatening hazard every winter.

Brief Transitions
Since there is a greater-than-even chance of having frost after May 25 and before September 20, the spring and fall seasons are of relatively short duration. Heavy

AVERAGE ANNUAL NUMBER OF DAYS WITH SEVERE WEATHER

	Thunder storms	Blowing Snow	Freezing Rain	Hall	Fog
Brandon	22	18	9	1	21
Winnipeg	27	25	12	3	20
Thompson		10		1	24
Dauphin	25	16	8		14
Churchill	7	64	19		48

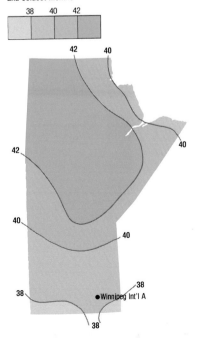

AVERAGE ANNUAL TEMPERATURE RANGE (°C)
Difference between average temperatures of warmest and coldest months

38	40	42

AVERAGE ANNUAL PRECIPITATION (mm)

450	500	550

● Per cent of annual precipitation occurring from May to September, inclusive

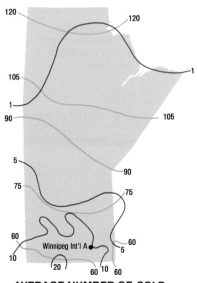

AVERAGE NUMBER OF COLD AND HOT DAYS

— Annual days with minimum temperature below −20°C
— Annual days with maximum temperature above 30°C

spring snowstorms can and do occur. On May 24, 1984, up to 30 cm of heavy, wet snow blanketed southwestern Manitoba, causing power outages and hazardous driving conditions. Spring is also the time of damaging floods, especially along the Red River and the Souris and Assiniboine rivers.

The fall season is also short. Any fair and dry weather after September and October is considered a plus. Heavy rains or wet snows can occur, but a persistent snowcover normally is not expected until November 20.

Manitoba is a windy province — open as it is to the free sweep of air currents from north and south. Winds are strongest in spring and fall. The strongest winds come from the northwest, but this direction may not be the most prevalent. North and northeast are the favoured directions in May and west in the fall. Strong spring winds are a problem to farmers. Sweeping unimpeded across freshly seeded but bare soil, the winds can cause severe erosion. May has the greatest average wind speed, 16 km/h, about 10 to 15% above the average for the year. July and August winds tend to be light. Churchill's strongest winds, like those elsewhere in the north, occur in the fall.

Weather Hazards

Each year Manitobans have to deal with the effects of severe weather — weather of exceptional intensity that interrupts normal activities and causes property damage and personal tragedy. In Manitoba, such

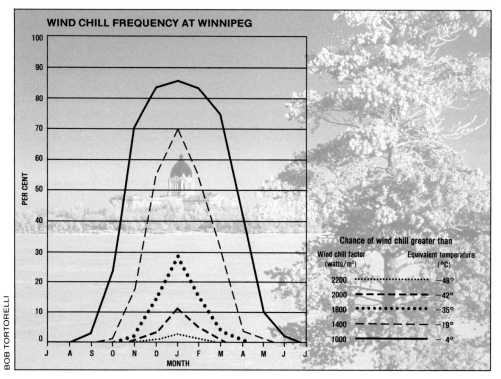

WIND CHILL FREQUENCY AT WINNIPEG

BOB TORTORELLI

PER CENT — MONTH (J A S O N D J F M A M J J)

Chance of wind chill greater than

Wind chill factor (watts/m²)		Equivalent temperature (°C)
2200	··········	−48°
2000	– – – –	−42°
1800	••••••••	−35°
1400	- - - -	−19°
1000	————	4°

TOTAL NUMBER OF TORNADO DAYS BY CENSUS AREA (1960-1982)

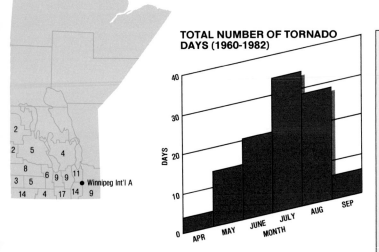

Winnipeg Int'l A

TOTAL NUMBER OF TORNADO DAYS (1960-1982)

DAYS — MONTH (APR MAY JUNE JULY AUG SEP)

MANITOBA WEATHER SUPERLATIVES

	What	Where	When
Temperature			
highest	44.4°C	Emerson	July 12, 1936
		St. Albans	July 11, 1936
lowest	-52.8°C	Norway House	Jan. 9, 1899
Greatest			
precipitation (mm)	966 mm	Peace Gardens	1975
snowfall (cm)	355 cm	Brochert	1960-61
number of fog days	70	Churchill	1975
wet period	195 days	Island Lake	1979
number of blowing snow days	84	Churchill	1960-61
sunshine	2706 hours	Delta University	1976
wind (maximum hourly)	130 km/h	George Island	Sept. 15, 1980
frost-free period	172 days	Vogar	1975

weather includes blizzards, ice storms, severe thunderstorms with lightning, and tornadoes.

BLIZZARDS

Blizzards are characterized by intense cold, strong winds, and low visibility in snow. In Manitoba blizzards, new snow is not always present; often most of the snow that piles into huge drifts has fallen earlier and is just redistributed during the storm. Blizzards and drifting snow are a hazard in the open prairie but less so in the forests where the force of the wind is broken by the trees.

One of the worst blizzards on record occurred on November 8, 1986, in Winnipeg. A total of 35.8 cm of snow fell on the city, with comparable or greater amounts throughout southern Manitoba. Winds gusting to over 90 km/h reduced visibility to zero at times and formed deep drifts in most areas. Travel was impossible, except for military vehicles, snowmobiles, and snowploughs. Snow removal costs approached $3 million, and it took up to a week for conditions to return to normal. The storm was similar to a blizzard in March 1966 that brought 35.6 cm of snow but was exceeded by one in March 1935 that was accompanied by snowfalls of 51.8 cm.

ICE STORMS

Freezing precipitation falls infrequently in Manitoba, but when it does occur it usually results in major disruption. Heavy icing, especially when accompanied by strong winds, devastates trees and brings down wires, disrupting communications and power supplies. Sidewalks and roadways become extremely hazardous, and ice loads on roads may become unbearable. One of the more vicious ice storms of recent years hit the province on March 6, 1983. Winnipeg airport was closed for two days, and several large TV towers collapsed under the weight of more than 28 mm of freezing rain.

Freezing rain and drizzle occur on an average of 10 to 15 days a year in southern Manitoba and 15 to 20 days a year along the shores of Hudson Bay. Winnipeg averages 30 hours a year of freezing rain, with December being the peak month and November a close second. Churchill has twice as many hours, with peak occurrences in November and May.

THUNDERSTORMS AND HAIL

Manitoba has many thunderstorms, particularly in the south where Winnipeg and Portage la Prairie average between 25 and 30 days a year when thunder is heard. The prime time for thunderstorms is the late afternoon and evening, between 5 and 10 p.m. The prime season is mid-June to late August, with a peak in the third week of July. This is the time when surface heating and vertical air movement are at their maximum. Thunderstorms accompanied by wind and hail cause local damage to crops, livestock, and property. Frontal thunderstorms that develop along cold fronts are especially destructive and affect large areas. About three times a year hail accompanies the thunderstorms. Most of the hail that falls is small. However, stones as large as golf balls are reported often in southern Manitoba during each summer, and stones as large as baseballs have occurred.

TORNADOES

Every year at least one tornado is reported in Manitoba. Because of the low density of population, loss of life is quite rare and damage is confined to farm buildings and telephone, power and transmission lines. Tornadoes in Manitoba favour July and August and the late afternoon and evening. The frequency is highest in the Red River Valley and adjacent areas.

Over the 22 years between 1960 and 1982, 134 confirmed tornadoes were reported, an average of six a year. On July 6, 1987, severe thunderstorms produced at least two tornadoes in southern Winnipeg. The winds caused considerable property damage, and more than 40 mm of rain in two and a half hours resulted in flash floods.

The tornado census points to a gradual increase in the numbers reported yearly. However, this is more likely a result of greater public awareness, increased population density, and a more extensive network of severe weather watchers than it is an indication of climate change.

A typical Manitoba tornado has a path 8 km long and 15 to 50 m wide. It strikes from the west, moving at 50 km/h and with maximum winds reaching 250 km/h.

PORTAGE AND MAIN - THE COLDEST CORNER IN CANADA?

Winnipeg—sometimes known as Winterpeg Manisnowba, Canada's winter city, or the city with two seasons, winter and July—has had to put up with many slurs about its legendary winters. So often has it been identified with cold weather that it has gained a kind of unofficial status as the cold capital of Canada.

The fact is, though, that determining which place has the coldest temperature depends largely on what is meant by coldest. Is it the coldest day, month, or winter? Is it the longest cold spell? Or is it the coldest on average over several years? Below are some Canadian cold superlatives. None of them, however, includes Winnipeg.

Coldest Day	-63°C	Snag Yukon	Feb. 3, 1947
Coldest Month	-47.9°C	Eureka, NWT	February 1979
Coldest Year	-21.7°C	Eureka, NWT	1972
Longest Cold Spell	223 days (< -18°C)	Isachsen, NWT	Oct.-May 1949-50

That cuts no ice with folks living in the south. But comparing Winnipeg's winter temperatures with those of other urban centres south of the Arctic Circle does not change the result. Yellowknife has the coldest midwinter temperature of centres over 5000 persons, and Thompson is the coldest for places over 10 000. Both places are colder than Winnipeg. For major cities over 100 000 population, Saskatoon is coldest, with a January mean daily minimum fractionally below that of Winnipeg.

Now combine bitterly cold temperatures with gusty winds. Winnipeg does fairly well in this department, with a wind chill record of 2622 W/m (equivalent to a still air temperature of -65°C) set on January 9, 1982. But again, Winnipeg cannot claim a record. Edmonton International Airport had a higher extreme on December 15, 1964—a numbing 2732 W/m (-69°C), the product of 55 km/h winds and an air temperature of -35.6°C. As for marathon bouts of severe wind chill weather (when the wind chill exceeds 1600 W/m i.e., an equivalent temperature of -25°C), Winnipeg's longest flesh-freezing episode lasted 170 hours, beginning on January 24, 1966. Saskatoon and Regina, however, have endured longer periods—216 and 214 hours respectively, beginning on December 28, 1978.

For Winnipeg, the city centre is usually 3 to 4°C warmer than the airport, owing to the urban heat island effect. But try telling this to the office worker struggling with one of the concourse doors of the James Richardson Building on the northeast corner of Portage and Main. There the funnelling winds may be 40% stronger than those observed at the airport. The city's accelerated winds, even with slightly warmer temperatures, combine to produce severe wind chills that are slightly higher downtown than at the open site at the airport.

Canada's Place Under the Sun

Travel ads, resort brochures, and chamber of commerce pamphlets often boast about abundant sunshine. With the help of lengthy records of sunshine measurements across Canada, the veracity of such claims can be put to the test. One conclusion that emerges clearly is that the Prairies are the sunniest region in Canada — almost all year-round.

As for which of the Prairie Provinces has the most sunshine, that distinction belongs to Saskatchewan, but only by a shade. Of the 320 stations in Canada with sunshine records of sufficient duration to provide a reliable average, Estevan is the sunniest with 2537 hours a year. Alberta's sunniest spot is Coronation with 2490 hours, and Manitoba has Gimli with 2392 hours. Saskatoon, with 2450 hours, is the sunniest major city in Canada and Regina the sunniest provincial capital with 2335 hours. The variation of sunshine across Saskatchewan and Alberta is relatively small. Prov-ince-wide, Saskatchewan averages about 50 hours of sunshine a year more than Alberta and is slightly above her neighbour in sunny hours in both winter and summer.

July is usually the sunniest month in Saskatchewan, and December is the dullest. April is sunnier than October and, in a few places, May is sunnier than June. In June, the wettest month, frequent storms cloud the skies and reduce the number of hours of sun, in spite of the 17-hour days that are possible at this time of year.

A redeeming feature of the Saskatchewan winter is the small number of sunless days, about 5 a month on average. In southern Alberta, on the other hand, winter often brings easterly winds which are generally associated with cloudy skies.

Climate Causes

Saskatchewan is in the heartland of North America, far from the influence of the

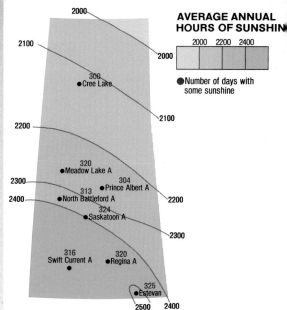

2000 2200 2400

●Number of days with
some sunshine

2000

2100

2000

309
●Cree Lake

2100

2200

2200

320
●Meadow Lake A

304
●Prince Albert A

313
●North Battleford A

2300
2400

2200

324
●Saskatoon A

2300

316
Swift Current A
●

320
●Regina A

325
●Estevan

2500   2400

SOME SUNNY CAPITAL CITIES

	Annual Sunny Hours
Regina	2331
Athens	2758
Kartoum (Sudan)	3924
London	1514
Madrid	2871
Moscow	1597
Ottawa	2009
Paris	1692
Rabat (Morocco)	3149
Rome	2491
Tokyo	2021

oceans and mountains. It experiences large daily, and extremely large annual, ranges of temperature — the greatest anywhere in the ten provinces. For a comparable climate in some other part of the world, one must look deep into the centre of Europe and Asia.

Saskatchewan's long winters are colder than Alberta's and on a par with Manitoba's. Its short summers are only 1 to 11/2 °C cooler than those in southern Ontario — some 800 km farther south. The province's mid-latitude position gives it about 17 hours of sunlight on the longest day of the year. In winter, 71/2 hours of sunshine are possible on the shortest day, December 21.

The province is a vast, gently sloping plain stretching from horizon to horizon, without any barriers to impede the movement of air masses. In the north, the Canadian Shield occupies one third of the province with a surface made up of thin soil, swamp, muskeg, lichen-covered rock, and forest. In

the south, the terrain is an undulating, hummocky plain that slopes from 900 m above sea level in the west to 300 m in the east. A number of minor upland areas of varying size rise above the surrounding plain but are ineffective as regional climate shapers. In the southwest, the province shares the Cypress Hills with Alberta.

Saskatchewan is too far away to enjoy the direct and immediate moderating effects of the Pacific Ocean, although occasional mild winter weather and much of its moisture comes to the province from the Pacific.

Saskatchewan is in the zone of the westerlies, with weather systems moving in a general west to east direction. It lies open to a variety of air masses — cool, moist Pacific; cold, dry Arctic; and warm, dry American continental — that meet and clash over the province with great frequency. Storms associated with these frontal clashes are present in all seasons of the year but are usually more pronounced in the spring. Moisture may come from light, general rains or from violent storms in which torrential rains fall for hours. Local showers and thunderstorms are common summer occurrences.

Dry Country

Saskatchewan is the driest province. Located as it is near the centre of the continent and shielded from the Pacific Ocean by the western mountains, Saskatchewan is usually dominated by air that is too dry to yield much precipitation.

Total yearly precipitation ranges from 300 to 500 mm or more. The heart of the dry country is the South Saskatchewan River basin along the Alberta – Saskatchewan border, where the scanty rainfall is made even less effective for agriculture by the strong summer sunshine and drying winds. Here the prairie is treeless and grass grows in short tufts, reflecting the paucity of

SASKATCHEWAN WEATHER SUPERLATIVES

	What	Where	When
Temperature			
highest	45.0°C*	Midale, Yellow Grass	July 5, 1937*
lowest	-56.7°C	Prince Albert	Feb. 1, 1893
Greatest			
precipitation	916 mm	Kamsack	1921
snowfall	368 cm	Pelly	1955-56
number of fog days	61	Collins Bay	1980
thunder days	54	Beechy	1977
wet period	180 days	Whitesand Dam	1978
number of blowing snow days	65	Moose Jaw	1973-74
sunshine (hours)	2701 hours	Estevan	1980
wind (maximun hourly)	142 km/h	Melfort, Agriculture Canada station. Oct 4, 1976
frost-free period	167 days	Aylesbury, Colonsay	1978

* Canadian record

KAY DIXON

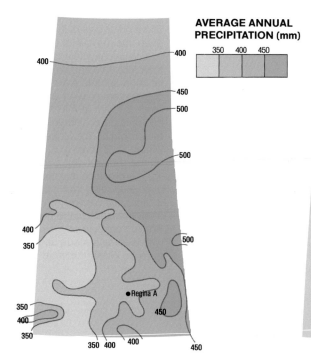

AVERAGE ANNUAL PRECIPITATION (mm)

350 | 400 | 450

400
450
500
500
400
350
500
● Regina A
350
450
350
400
350 400 400 450

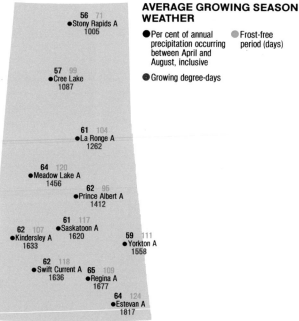

AVERAGE GROWING SEASON WEATHER

● Per cent of annual precipitation occurring between April and August, inclusive

● Frost-free period (days)

● Growing degree-days

56 71
● Stony Rapids A
1005

57 99
● Cree Lake
1087

61 104
● La Ronge A
1262

64 120
● Meadow Lake A
1456

62 95
● Prince Albert A
1412

61 117
● Saskatoon A
1620

62 107
● Kindersley A
1633

59 111
● Yorkton A
1558

62 118
● Swift Current A
1636

65 109
● Regina A
1677

64 124
● Estevan A
1817

moisture in this dry belt.

Fortunately, the bulk of the province's modest precipitation is strongly concentrated in the growing season from May to August. In June the primary storm track is positioned across the centre of the province. Between June and July it shifts north of the Prairies. As a result, June is usually the wettest month, with rainfalls of 60 to 70 mm on average. Pacific air brings frequent summer showers, with rain falling 10 to 15 days a month. These showers are usually associated with cold fronts or summer thunderstorms. In some years warm, moist tropical air moves northward from the Gulf of Mexico and relatively heavy rains occur. Southeastern Saskatchewan benefits the most from these extensive systems that bring "million-dollar" rains to thirsty grain fields. One particularly wet period occurred in August 1966 near Wynyard, when it rained steadily for 78 hours, with total amounts around 150 mm.

September and October are noticeably drier, a definite advantage at harvest time. The foregoing describes an average rainfall pattern. In any one year, precipitation may be less or considerably more. Drought may occur one season and floods the next. Irregular, essentially unreliable precipitation over space and time is a characteristic of the Saskatchewan climate.

Moisture Deficits

High temperature, low humidity, strong winds, and clear skies lead to enormous evaporative water losses. In southern Saskatchewan, the need of growing plants for water far exceeds the amount of water available from rainfall and stored soil moisture. In most years the demand/supply situation creates a moisture deficiency of between 150 and 250 mm across southern Saskatchewan. At Swift Current the average annual precipitation is about 380 mm, whereas the water loss from an evaporation pan during the growing season alone is about 822 mm.

Snowfall

Snowfall is important to Saskatchewan. By filling water courses and replenishing soil moisture and groundwater reserves, it plays a crucial role in maintaining adequate water supplies for agriculture, hydroelectric, and domestic consumption. Snowcover prevents soil from drying out and prevents soil erosion. It also protects root systems from cold damage. But it can cause problems as well. A deep snowpack restricts foraging, and ice crusts which form over the snow may injure livestock.

Snowfall is not heavy, the annual total fall in the southern part of the province being about 100 cm. Regina's average yearly

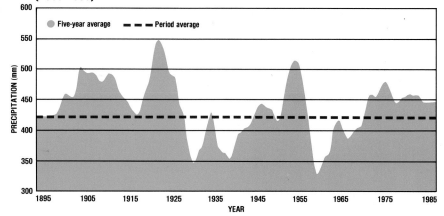

ANNUAL PRECIPITATION (mm) AT INDIAN HEAD (1895-1985)

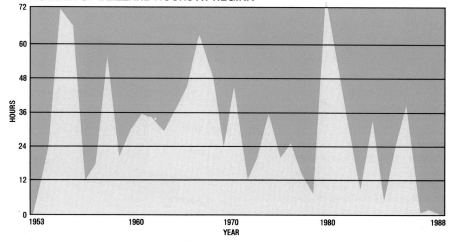

NUMBER OF BLIZZARD HOURS AT REGINA

snowfall of 116 cm is the same as that of Windsor, Ontario, Canada's most southerly city. In the north, however, snowfall amounts increase to 175 cm.

Snow may fall in September and May, but generally it falls only from November to April in most of the agricultural region. In the far north, the snowcover is present for over half the year. On the shores of Lake Athabasca, there are on the average 185 days with snow.

Violent Summer Weather

TORNADOES

Tornadoes are much less frequent and not nearly as intense in Saskatchewan as they are in the central United States. During the period 1960 through 1982, there were 211 confirmed tornadoes in the province. These occurred on an average of 7 days each year. Southeastern Saskatchewan is the most active region, although tornadoes have been sighted as far north as Deschambault Lake, east of La Ronge. July is the preferred month for tornado activity, and 4 to 7 p.m. is the most common time period. The

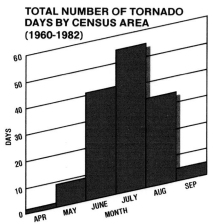

TOTAL NUMBER OF TORNADO DAYS BY CENSUS AREA (1960-1982)

KAY DIXON

tornado season extends for 100 days, from the third week in May to the last week in August.

Although tornadoes are small and rarely cause any widespread damage or loss of life, it is not unusual for certain places to be struck again and again. Regina has been hit on at least five occasions, with three hits in 1979, in May, June, and August.

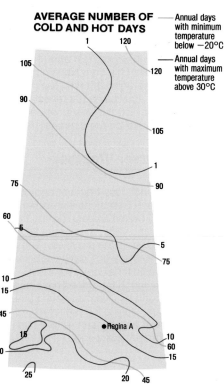

AVERAGE NUMBER OF COLD AND HOT DAYS

— Annual days with minimum temperature below −20°C

— Annual days with maximum temperature above 30°C

Thunderstorms and Hail

Saskatchewan has its share of violent thunderstorms, particularly in the southeastern part of the province where an average of 25 thunderstorm days occur each year. July is the month of peak activity, with thunderstorms occurring on an average of 7 days or 11 hours during the month.

There is seldom a summer day without a thundershower somewhere in southern Saskatchewan. Rain accompanies about half of the storms, and while rainfall does not last long, it is often quite intense. Rainfall rates of 20 mm per hour are typical. One storm dumped more than 250 mm of rain within

one hour at Buffalo Gap in south-central Saskatchewan on May 30, 1961, setting a Canadian record. Road washouts and field erosion were extensive as a result of this storm. All vegetation was removed from some crop and pasture areas; bark was even stripped from several trees; and paint removed from grain elevators.

Light falls of hail accompany many Sas-

TOTAL NUMBER OF TORNADO DAYS BY CENSUS AREA (1960-1982)

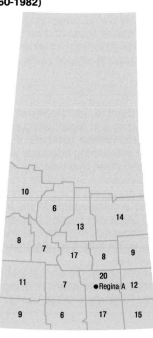

METEOROLOGICAL MOMENT

It was June 30, 1912, and Regina was readying itself for the next day's Dominion Day celebration. Flags and banners hung listlessly in the heavy Prairie air. It had been a blistering four days, with afternoon highs above 30°C. Most Reginians were just trying to keep cool, fanning themselves on verandahs or paddling canoes on Wascana Lake.

At about 4:50 p.m., dark sinister clouds appeared to the southwest, 4 to 5 km out of the city. While the clouds approached, they formed a funnel. Within minutes the whirlwind swept through Regina, slashing a path of destruction six blocks wide and tearing down about 500 buildings. When it was over, the Regina cyclone had become the worst killer tornado in Canadian history, leaving 28 dead and hundreds injured and causing $75 million damage (1990 dollars).

katchewan thunderstorms, but each summer, on the average, 10 to 15 severe hailstorms inflict considerable damage. Damage varies with the intensity and the size of the hailstones — most are the size of small pellets, but on occasion golf ball to baseball-size stones fall. Light hail is very common and almost non-destructive. On the other hand, hail with strong winds can be very destructive and can level a thousand square kilometres of standing crop in a bad summer.

Hailstorms may occur anywhere in the province but are most frequent in the dry lands to the southwest, where frequencies are not unlike those in "hailstorm alley" between Edmonton and Calgary. Hailstorms are most prevalent in Saskatchewan during the early summer. About three quarters of the hailfalls occur between noon and 8 p.m., and a typical hailfall lasts about six minutes.

Prairie Blows

Saskatchewan is not the windiest region in Canada; that distinction is shared by the coasts. However, open as it is to the free sweep of winds from any direction, the province does experience strong average wind speeds. Because the terrain is relatively flat and the land is largely cropped or in pasture, obstacles that impede the wind are lacking. For this reason winds are generally lighter in the more northerly treed parkland than in the open prairie.

In winter, northwest winds predominate across the province, with much stronger winds in the south than in the north. The average January wind speed at Swift Current is 25.1 km/h, but it is only 8.6 km/h at Uranium City on the shores of Lake Athabasca. In the spring, southeasterly winds predominate. April and May are often the months with the greatest wind speeds, averaging from 5 to 15% above the mean for the year.

Southerly winds in summer come from the American southwest and are generally hot and dry. They waste much valuable moisture, damage crop foliage, and carry away productive top soil. Each year in the south, an average of 15 hours of dust or sand is blown with sufficient force to reduce visibility. Occurrences of blowing dust or sand peak in the spring: 70% of the observations occur in April and May, at the time of highest wind velocity and when the vegetative cover is sparse and the topsoil layer is usually dry.

During the drought of the 1930s, wind erosion of soil produced massive dust-storms. These black blizzards, as they were called, caused widespread crop destruction, farm abandonment, and wholesale migration. Modern methods of soil erosion control — shelter belts, fencing, conservation tillage, strip cropping, fall-seeded crops, and crop rotation — have been effective in

minimizing soil losses but have not eliminated the threat completely.

Cold

BUT ALSO CANADA'S HOTTEST

Saskatchewan's winters are cold, but its summers are hot. The coldest day of the year at Regina is January 17, with a mean daily temperature of -18.7°C. The warmest day is July 21, with a mean of 19.3°C.

Less than half the year is free from freezing temperatures, but there is an 84% chance of having what is sometimes called a "bonspiel" thaw in January at Regina. Daily temperature records in the farming area of the province show that it is also possible for frost to occur at almost any time of the year. Saskatchewan's extreme temperature range extends from -56.7 to 45°C. Within this span of more than one hundred degrees, there have been January temperatures well above freezing and July temperatures below.

In winter the cold is not only severe but of extremely long duration. A cold spell may last for several weeks, as one did, for example, in the winter of 1955-56 when the temperature at several stations in southern Saskatchewan remained below -10°C for a period of 129 days.

During the summer there is generally less variation between the northern and southern parts of the province. The warmest region, around Estevan, has an average July temperature of 19.9°C, while at Uranium City on the shores of Lake Athabasca the average is 16.2°C, only 3.7°C cooler. In contrast, the January difference between mean temperatures at these two locations is 11°C. However, temperature differences between night and day are greater in summer than in winter; the average spread between night and day is

14°C in summer and 10°C in winter.

On hot summer days temperatures can rise to 40°C. In fact, the highest temperature ever recorded in Canada, 45°C, occurred in Saskatchewan, at Midale and Yellow Grass, on July 5, 1937. The previous July had also been marked by exceptionally hot weather, with daily highs reaching or surpassing the 40° mark on several occasions. During this period, daily temperatures averaged out to between 21 and 24°C at a number of stations in Saskatchewan and east into Ontario, making July 1936 the warmest month on record in Saskatchewan and Canada.

But hot days are not the whole story. Saskatchewan summers are also marked by low humidity and cool nights, resulting in pleasant weather and ideal sleeping conditions.

SAMPLE 1989 HAIL RATES FOR TOWNSHIPS IN SOUTH EASTERN SASKATCHEWAN

Rates are converted to cash rates and multiplied by a surcharge depending on the crop type (Great American Insurance Company). Rates are based on history of hail occurrences. Note wide variability between adjacent townships.

4	4.5	2.5	2.5	3.5	2.5	3	3	4	2.5	2	2
2	2.5	3	2.5	3	3	3.5	3	4	2.5	2	2
2	2	2	3	3	3	3	3	3.5	3	5	4
3	3	3	2	2	3	4	3	4.5	4.5	6.5	9.5
3	4	2.5	2	2	3	4	4	4	7	8	10
5	4	3.5	2	2	3.5	4.5	4.5	5.5	5	7	8
6	5.5	3	2.5	3	3	3	4.5	5	5	5	5
5.5	4.5	3.5	4	4	3	4.5	6	5.5	6	6	7
3.5	4.5	4.5	5	5	4	7	9	8	8	6.5	6
4.5	3	3.5	3	3	4.5	7	8	9	8	6	5
5.5	5.5	2.5	2.5	4	5.5	3	7.5	7	5.5	5	4
7	6	3	3.5	3.5	2.5	3	5	6	5	4	3

• Saskatoon A

• Regina A

• Weyburn
• Estevan A

NUMBER OF HOURS OF BLOWING DUST AND/OR SAND AT MOOSE JAW 1954 TO 1985

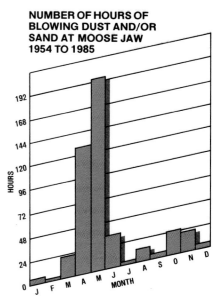

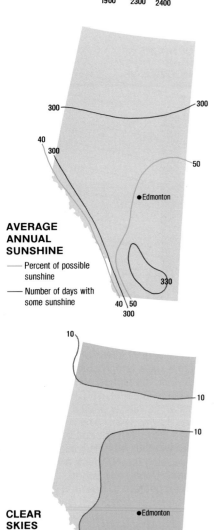

AVERAGE TOTAL HOURS OF SUNSHINE

2100
2000
1900
2200
2300
2400
2100
● Edmonton
1900 2300 2400

2000	2200	2400

AVERAGE ANNUAL SUNSHINE

300 — 300
40
300
50
● Edmonton
330
40 \ 50
300

— Percent of possible sunshine
— Number of days with some sunshine

CLEAR SKIES IN JULY

10
10
10
● Edmonton
20
30
10 20 \ 30

Per cent frequency of clear skies in the afternoon

10	20	30

LAND OF CHINOOKS AND BIG SKIES

JOAN SCHWARTZ

Albertans have come to expect the unexpected in their weather and climate. In any year, flowers may bloom in February, snow may fall in August, and chinooks can turn winter into spring in an hour. No two years are alike. Drought spells one year give way to snowmelt floods the next. At one time dinosaurs roamed the badlands, and forests thrived in tropical climes, and at others, ice lay a kilometre thick across much of the province.

Today Alberta is home to snow-eaters, snow rollers, dust devils, and the rain-shadow. It shares a continental climate with its Prairie neighbours but is generally warmer and drier and in some circumstances sunnier than Saskatchewan and Manitoba.

Big Sky Country

Alberta boasts some of the clearest, driest, bluest, and sunniest skies in Canada. Arctic air is usually dry and clear, while mild-Pacific air arrives on the plains wrung dry and nearly cloud-free after its trek over the western mountains.

Alberta competes closely with Saskatchewan for the title of sushine province of Canada. Annual bright sunshine totals exceed 2000 hours almost everywhere. The provincial high, at Suffield and Coronation, is nearly 2500 hours. Summer is the sunniest season, in part because in the northern part of the province summer days have up to 18 hours of daylight. During an average summer 60% of the daylight hours have bright sunshine, and there are only three days when the sun is entirely absent. In October skies range from clear to only partly cloudy half the time, making this the month with the highest percentage of hours with clear skies. About 40% of the total possible hours in winter have sunshine, one of the highest "percentage-of-possible" totals anywhere in southern Canada.

Alberta has some of the driest air in the country. In July the average vapour pressure (the best measure of humidity) is 1.1 kPa — far below the 1.6 to 1.8 kPa averages in southern Ontario. As far as is known, Calgary had the lowest relative humidity, 6%, ever recorded in Canada. This figure was reached on March 22, 1968, when the air temperature was 18°C and the dew-point -20°C.

Alberta is also one of the least foggy areas of Canada. In the southern and western portions of the province, there are 12 to 15 days with an hour or more of fog each year. In the northwest, a few additional ice fog days push the province-wide average of 20 days a year to 30 days. Some low-lying, polluted areas are prone to a few extra days with fog each year.

Making Alberta's Climate

Alberta has a continental climate with long, cold winters, and short, cool summers, and low yearly rainfall. It lies in the same belt of latitude as damp and cold Labrador, maritime Ireland, the steppelands of Eurasia, the deserts of northern Mongolia, and the North China Plain. But latitude alone does not determine climate. Altitude, distance from the oceans, the prevailing circulation of the atmosphere, and local geographical features collectively shape climate. This is nowhere truer than in Alberta.

TOPOGRAPHY AND CIRCULATION

The Rocky Mountains form an effective barrier to the maritime influence of the Pacific Ocean. Cool North Pacific air loses considerable moisture coming over the mountains and is warmed while it descends the eastern slopes of the Rockies and arrives on the rain-shaded Prairies. Still, air of Pacific origin gives rather cloudy, mild, and windy weather to Alberta. Precipitation can be relatively heavy in the foothills and in the Peace River district, where altitude decreases and rain-bearing air masses enter the province freely from the west. However, nowhere is the yearly total precipitation excessive. By way of comparison, Montréal's yearly total of 1070 mm exceeds that of any Alberta station.

With no east-west mountain chain over the region to separate contrasting air masses, such as the Alps in Europe and the Himalayas in Asia, cold, dry Arctic air and warm, dry air from the American southwest meet and clash with great frequency over Alberta. Consequently, Alberta's continental climate features a wide variation of temperature, cloud, and precipitation between summer and winter and from day to day in every season.

OTHER CONTROLS

Alberta rarely experiences the moderating effects of the Atlantic Ocean or of the Gulf of Mexico. The Pacific Ocean, however, is less than 500 km away, and it provides some moderation in the form of milder winter temperatures compared with those experienced in Saskatchewan.

Local geographical and topographical features have important climatic effects. Tops of hills are likely to be colder than their slopes. On the other hand, cold air drains downhill to valley floors, where frost may occur prematurely on clear, cold, still nights. Lakes and major rivers may add enough warmth to nearby areas in summer and fall to prevent frosts and allow the growing of cool-season crops, especially in the northern and central parts of the province. Finally, Alberta's large cities significantly alter their climates by warming air temperatures and by providing shelter from the winds.

Running Hot and Cold

Of all the Prairie regions, southwestern Alberta has the smallest annual temperature range, 26°C. (That's the difference between the average temperature of the coldest winter month and the warmest summer month.) This compares with a range of 43°C for the northern portions of all three provinces. On the other hand, southwestern Alberta is also famous for its chinook winds, which can cause temperature swings of 20°C in less than an hour.

Winter cold across Alberta is less severe than in the two neighbouring Prairie provinces. Several stations in southwestern Alberta have average January temperatures around -10°C, about 8°C higher than in Saskatchewan and Manitoba. In the far north all three provinces have average winter temperatures of -25°C for the coldest month.

Two characteristics of Alberta's winter temperatures are extreme variability from hour to hour and from year to year and lengthy spells of persistent extremes. The chinook and its sudden shifts have already been noted. The coldest winter on record in southern Alberta occurred in 1968-69 when the December to February average temperature was -16°C, compared to a normal of -7.2°C. At Edmonton the winter of 1886/87 was the coldest, with a December to February average of -20.0°C, 7°C below normal. Only two winters later, Edmonton experienced its mildest winter ever, with temperatures 6°C above normal. In 1911 Fort Vermilion recorded the all-time provincial low of -61.1°C, a reading that remains unchallenged outside Yukon as the coldest ever observed in Canada. In some winters, under persistent Arctic air, a cold spell may last for weeks, as, for example, in 1969 when Edmontonians experienced 26 straight days when the temperature failed to rise above -18°C (0°F).

In some years, surprise May snowfalls are followed by the year's first heat wave, and October's prolonged Indian Summer ends abruptly with the winter's first snowfall or killing frost. In almost every year spring and fall are the shortest seasons, lasting only about a month.

During the summer, there is generally less temperature variation between northern and southern areas of the province than at other times of the year. Fort Smith, just

HANS VANLEEUWEN

GOOD WEATHER DEFINITION

Activity	Temp. °C	Visibility (km)	Hourly Wind (km/h)	Snow Cover (cm)	Precipitation
Sightseeing	—24 to 32	›5	‹42	—	nil
Skiing or Snowmobiling	›—14	›1	‹26	›3	nil or light
Beaching	›18	›1	‹26	—	nil

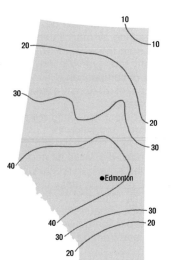

PER CENT CHANCE OF GOOD SKIING OR SNOWMOBILING WEATHER IN MID-JANUARY

PER CENT CHANCE OF GOOD SIGHTSEEING WEATHER IN MID-JULY

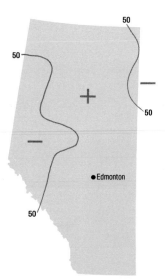

PER CENT CHANCE OF GOOD BEACHING WEATHER IN MID-JULY

over the border in the Northwest Territories,is one of the coolest places, with a July mean temperature of 15.7°C, while at Milk River, near the U.S. border, it is only 3°C warmer. Excluded from this comparison are the snowcapped mountainous peaks in the west where typical July averages run between 10 and 14°C.

Frost

Southeastern Alberta, near Medicine Hat,has the longest frost-free period in the province and on the Prairies, 125 days. North and west of Medicine Hat and Lethbridge, the average date of the last spring frost is delayed from late May to mid-June, whereas the average date of the first fall frost advances from mid-September to late August. Most grain-growing areas in the province, including the Peace River district, have between 100 and 120 frost-free days, which is enough for modern strains of grain to mature. However, the limiting factor agriculturally is the "killing" frost, defined as -2°C or lower. The killing frost-free season extends about 20 to 25 days longer than the frost-free season in Alberta.

Once again these data do not apply to the rugged terrain of the Rockies, where the frost-free season varies dramatically over a short distance. The same qualification applies to other places where local landscape features, such as frost hollows, lakes, or dense urbanization, can affect temperatures. Edmonton city, for example, has an average 140 day frost-free season, compared with 105 days south of the city in the rural setting of the international airport.

Dry Country

Some of the driest parts of Canada are in southern Alberta. The province lacks available sources for abundant precipitation. Arctic air is dry, and moister tropical air rarely enters the province. As previously mentioned, air from the Pacific is wrung dry in its ascent of the western mountains. Most eastward disturbances that do make it into the plains are generally weak and still drying as they descend the eastward sloping land.

It is rare to have a summer day without a

shower somewhere over southern and central Alberta. Thunderstorms are often a source of welcome rainfall, but they may also bring unwelcome falls of hail.

As little as 300 mm of moisture falls yearly in the "dry belt" of southeastern Alberta — an area also distinguished for its rainfall variability and for its prolonged droughts. All of Alberta's agriculture is conducted in areas with under 500 mm of precipitation annually. Fortunately for the province's farmers, the rainy season corresponds with the growing season, from late May to early September. June, the wettest month, is also the time when the crops can best use the moisture. June rain totals are in the 75 to 100 mm range, but by July and August, storm tracks generally shift northward and monthly precipitation drops below 75 mm. Unfortunately, however, this pattern cannot always be relied upon. As a result, agriculture in Alberta is a high-risk business. Rainfall may be in short supply one season and excessive in the next. To alleviate the uncertainty, irrigation is extensively practised.

Precipitation also remains light outside the agricultural areas. In the north, annual totals are under 400 mm, while in the passes through the Rockies totals exceed 600 mm — with most falling not in June but in December. About 60% of the precipitation in the mountain passes falls as snow.

Prairie Legends

Except for the foothills region, snowfall is not heavy anywhere in Alberta. The yearly total is normally 150 cm everywhere — about 100 cm less than that which normally falls between Ottawa and Montréal. The area near Lloydminster on the Alberta — Saskatchewan border receives the least

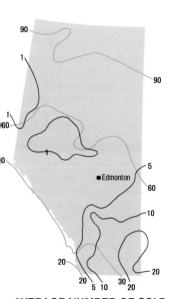

AVERAGE NUMBER OF COLD AND HOT DAYS

— Annual days with minimum temperature below −20°C

— Annual days with maximum temperature above 30°C

AVERAGE ANNUAL SNOWFALL (cm)

100 200 400

● Snowcover season (days) when snow on ground exceeds 2 cm

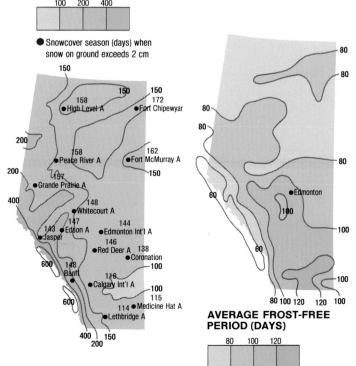

150

150   150
158   172
● High Level A   ● Fort Chipewyar

200

158   162
● Peace River A   ● Fort McMurray A

157
● Grande Prairie A   150

148
● Whitecourt A
147   144
143 ● Edson A   ● Edmonton Int'l A
● Jasper   146
● Red Deer A   138
● Coronation
148   100
Banff
118
● Calgary Int'l A   100
115
114 ● Medicine Hat A
● Lethbridge A

AVERAGE FROST-FREE PERIOD (DAYS)

80 100 120

AVERAGE FROST-FREE PERIOD (DAYS) map labels:
80
80
80
● Edmonton
100
60
60
100
80 100 120   120 100

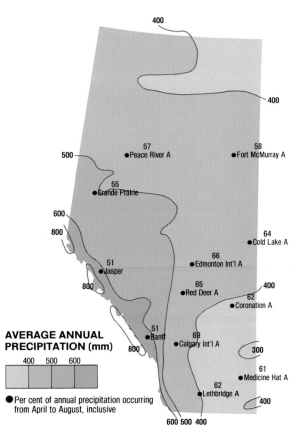

AVERAGE ANNUAL PRECIPITATION (mm)

400 500 600

● Per cent of annual precipitation occurring from April to August, inclusive

amount of snow, less than 100 cm. The Rocky Mountain ski resort of Lake Louise receives the most, more than 400 cm. In usually dry Alberta, snowfall accounts for 30 to 35% of the annual precipitation. The unusual thing about snowfall in Alberta is that records at most stations show an occurrence of snowfall in every month of the year — often just traces in summer but snowfall nevertheless.

Snowcover is generally at its maximum over the southern half of Alberta in February, but is usually less than 20 cm. The first snow appears in late October, and the last snow melts by May 1. A persistent snow- cover can be expected between mid-November and late March. The snowcover season is longer in the northern forests, but in chinook country the ground is often laid bare several times a winter.

Blizzards are as much a part of the Alberta climate story as the chinook. They are characterized by intense cold, strong winds, and snow, with visibility reduced below 1 km in blowing snow. The amount of snow may be very light, but, driven by fierce winds, it can cripple transportation and endanger life. February is the month when the storms are most likely to occur. Spring storms often do not meet one or two of the criteria to qualify as a full-fledged blizzard. But in many ways, these spring surprises are more disruptive to an unsuspecting populace. A storm on May 14, 1986, was a classic case, with knee-deep snow pushed around by 80 km/h winds. Though technically not a blizzard (it lacked the intense cold) this storm nevertheless affected more than a million southern Albertans, stranding motorists and leaving dozens of communities without utility services.

MYRLE CHYNOWETH

Big Blows

Winds blow unimpeded across the grasslands of Alberta — open as they are to the free sweep of air from several directions. In winter, southwesterly and westerly winds predominate, especially in the south where chinook winds can reach speeds of 100 km/h. In the north there are usually more winter calms than winds from any single direction. In the other seasons, west-northwesterly winds predominate. The average wind speed is strongest in spring, averaging from 10 to 15% above the yearly average of 13 km/h.

Thunderheads and Hailstorms

On many summer afternoons billowing clouds often develop into massive thunderheads. Some produce insignificant showers that do little more than wet the ground. Others unleash torrents of rain and an incessant barrage of hailstones — some the size of gun-shot and some as large as golf balls or baseballs that tear holes in roofs, dent automobiles, and flatten crops.

Thunderstorms occur an average of 20 to 25 days each year across southern and central Alberta, and between 10 and 20 days in the north and forested foothill regions. July has an average of 5 thunderstorm days throughout most of southern Alberta.

Central Alberta between Drumheller and the Red Deer — Calgary corridor has some of the most numerous and severe hailstorms in the world. Some farms have been hit as many as ten times in one year. Outside this "hailstorm alley," hail is not as frequent but may be equally damaging. Hailstorms may last for several hours and cut a swath 50 km wide and hundreds of kilometres long. July is the leading hail month in Alberta with 21 days; 6 to 9 p.m. are the peak hours. The average duration of hailfall for any given location is 10 minutes. By September, thunderstorm activity has dropped to less than 1 day a

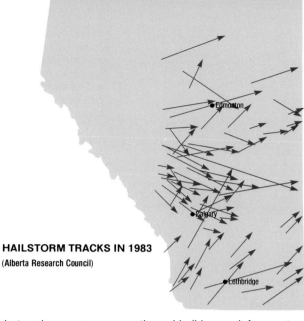

HAILSTORM TRACKS IN 1983
(Alberta Research Council)

month, and hail is very infrequent.

On rare occasions, brief but violent tornadoes drop out of the thunderhead. They are seldom as frequent, as intense, or as long-lived as those in southern Manitoba or southwestern Ontario. In one 30-year period, there were reports of 169 tornadoes on 118 days. In total, 34 fatalities occurred. Only a few of the twisters were considered severe or devastating.

Drought

THE EVER-PRESENT MENACE

In semi-arid Alberta, drought is an old and continuing problem, perhaps the greatest weather menace of all. Recurrent drought in the late 1800s and again in the 1930s led settlers to abandon their farms. Long-term records of annual precipitation show extreme year-to-year variability. Even small departures from normal rainfall will cause concern, and the situation can become catastrophic when two, three, or more dry years occur in succession. Prolonged droughts bring total crop failure and serious soil erosion, not to mention family and community disruption and regional economic collapse.

Alberta Snow Eaters

At midnight on January 27, 1962, the temperature was a frigid -19°C at Pincher Creek, Alberta. The weather was clear, winds were light from the east, and the relative humidity was 83%. Within an hour, a dramatic change took place — the temperature jumped 22°C and the winds swung around to west-southwest, increasing to 68 km/h with gusts to 95 km/h. Although the skies stayed clear, the relative humidity dropped to 56% and the snowcover disappeared like magic. Although spectacular, the weather change was not unusual. Resi-

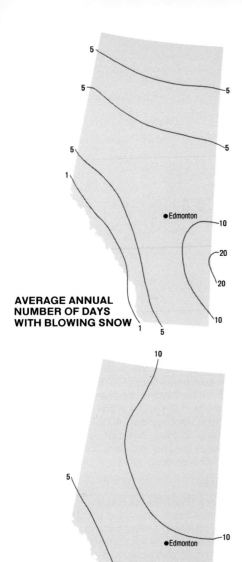

AVERAGE ANNUAL NUMBER OF DAYS WITH BLOWING SNOW

AVERAGE ANNUAL NUMBER OF DAYS WITH FREEZING PRECIPITATION

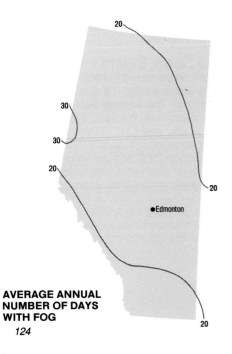

AVERAGE ANNUAL NUMBER OF DAYS WITH FOG

dents knew a chinook was under way and that the weather might continue springlike for days or revert just as suddenly to its previous calm and frosty state.

Chinook is an Indian word meaning "snow-eater." It refers to a class of strong winds called foehn winds that become warm and very dry in descending a mountain slope. In other mountainous regions the foehn (the name originated in the Alps) has a variety of local names: *zonda* in Argentina, *koembang* in Java, and *puelche* in the Andes.

Chinooks occur often over the western plains of North America but are strongest in southwestern Alberta, where they funnel through the Crowsnest Pass and fan out across southern Alberta, diminishing in effect as they cross into Saskatchewan. Chinooks occur on an average of 30 days per winter in the Crowsnest Pass, 25 days in Calgary, 20 days at Medicine Hat, and only 10 days at Swift Current. Edmonton is not prime chinook country.

Signalling the arrival of a chinook wind is the familiar arch of cloud in the western sky parallelling the Rockies. As it passes through, temperatures rise with startling rapidity, and the air becomes as dry as that of the Sahara. Chinook winds always blow

ALBERTA WEATHER SUPERLATIVES			
	What	Where	When
Temperature			
highest	43.3°C	Fort Macleod	July 18, 1941
lowest	-61.1°C	Fort Vermilion (Agriculture Canada station)	Jan. 11, 1911
Greatest			
precipitation	1440 mm	Cameron Falls	1975
snowfall	1066 cm	Columbia Ice Fields	1973-4
number of fog days	65	Whitecourt	1972
wet period	186 days	Fort McMurray	1972
number of blowing snow days	33	Coronation	1973-74
sunshine	2785 hours	Manyberries	1976
wind (maximum hourly)	108 km/h	Bighorn Dam	Jan. 15, 1972
frost-free period	184 days	Edmonton Municipal	1980

from the southwest or west and are strong (40 to 80 km/h) and gusty (up to 160 km/h), enough to rip roofs, snap power lines, shatter windows, and topple pedestrians.

What causes the chinook? The process begins with westerly winds carrying warm, moist air from the Pacific. When they reach the Coast Mountains, they glide up and over the western slopes, expanding and cooling as they rise. At higher elevations the air cools about 1°C for every 200 m of rise. In the ascent the water vapour condenses and falls as rain or snow. At the same time, great quantities of stored heat are released into the atmosphere. The process repeats with each successive interior mountain chain traversed, and with each ascent more moisture is wrung from the eastwardly travelling air stream. Once over the final mountain barrier, the Rockies, the now dry air plunges to the plains, warming by compression (like air warmed in a bicycle pump) at a rate twice the cooling rate of the ascent. Compression of the air coming into Alberta raises the temperature 8 to 10°C above that of the Pacific air at the same elevation on the west side of the Coast Mountains.

To southern Albertans, the chinook wind is a mixed blessing. Although to most, it may bring a welcome touch of spring and temporary relief from the bitter cold of winter, a few are bothered by debilitating reactions ranging from sleeplessness to severe migraines. Other drawbacks are decreased soil moisture, wind-eroded dry soil, and disrupted winter sports. But there are benefits too. Starving cattle and wildlife, separated from their food by heavy snow cover, have been saved by timely chinooks. Natural snow removal is another bonus. And, of course, there are the blessings of golf in January and lilacs before spring.

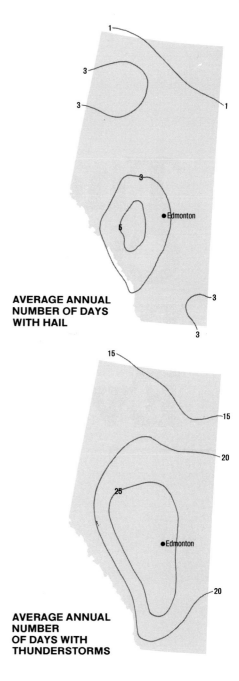

AVERAGE ANNUAL NUMBER OF DAYS WITH HAIL

AVERAGE ANNUAL NUMBER OF DAYS WITH THUNDERSTORMS

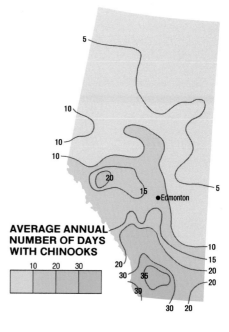

AVERAGE ANNUAL NUMBER OF DAYS WITH CHINOOKS

10   20   30

ALBERTA SNOW-EATERS A TALL TAIL

A farmer and his wife drove to town one winter day and a chinook blew up. The farmer put his wife in the back of the sleigh and started for home at full gallop. Keeping his team racing as hard as he could, he kept the front of the sleigh on the snow, but the rear runners dragged in the mud. By the time they got home, the farmer was frozen and his wife had sunstroke.

The Climate With Everything

The almost infinite variety of Canada's climate is nowhere more striking than in British Columbia. Mighty rivers and parched valley bottoms; Mediterranean mildness and wet, foggy coasts; dripping Pacific rain forests and cold, snowy plains; perpetually snow-capped mountains and balmy Gulf Islands — these climatic landscapes and many more, can be found in British Columbia. Within the Greater Vancouver area it is possible to play golf and ski on the same midwinter day. In a few hours drive along the spectacular Coquihalla Highway one can move from lush rain forest to semi-arid sagebrush country. In fact, British Columbia is the wettest and almost the driest place in Canada. It also comes close to being the hottest and coldest. On July 16, 1941, Lytton, Lillooet, and Chinook Cove recorded a temperature of 44.4°C, Canada's second highest, while to the north, Smith River recorded a temperature of -58.9°C on January 31, 1947, lower than any reading observed outside of Yukon and Alberta.

Its geography explains many of the unique features of the climate. British Columbia encompasses a latitude spread of nearly 1300 km with elevations from sea-level to 4663 m (Mount Fairweather). Watered by the relatively warm waters of the North Pacific and separated from the continental interior by a series of mountain ranges paralleling the coast, the province features a number of climate regimes. The major contrast is between the coast and the interior, but there are also significant variations between valleys and highlands, and between north and south.

Pacific Coast...

ALMOST MEDITERRANEAN

The Pacific climate region consists of the islands and a thin coastal strip of west-facing slopes, uplands, and indented fiords that extend no more than 150 km from the sea. Pacific air streams ensure mild winters, mild but not hot summers, and small seasonal temperature differences, making for the most consistent and moderate climate in Canada.

Temperatures stay above freezing in winter but seldom rise above 30°C in summer. Although it is not guaranteed, a wave of Arctic air usually floods through the valleys of the Coast Mountains once or twice each winter, engulfing the coast in

METEOROLOGICAL MOMENT

Prince Charles and Princess Diana opened Expo '86 on May 2 amidst rain, overcast skies, and cool temperatures. Fortunately, that was not the weather Expo was to be remembered for. On the contrary, fair-goers enjoyed a sustained spell of dry and sunny conditions that surpassed anything Expo's organizers could have hoped for.

From May 2 to October 13, the period when Expo operated, 253 mm of rain fell, just under the average of 269 mm at Vancouver International Airport. What was significant, though, was the timing: it rained on only 36 of the 165 days of the fair, and most of that fell at night. That means 78% of the days, or almost four out of five, were rainless. During Expo there was a 53-day dry spell from July 18 to September 8 and a 13-day dry spell on the last days of the exposition. September was seasonably mild and dry. Only the 23rd, with 48 mm of rain, was a washout, but that took care of 67% of September's total rainfall. October was dry and sunny too, and eager crowds flocked to the fair until the very end.

When it was all over, some 22 million people had attended. The fair's organizers, in their more optimistic moments, had hoped for 13.8 million, but then they hadn't counted on the weather.

frigid air. As a result, deep snows are not a rarity in Vancouver and Victoria, although rain is more common and there have been some winters with no snowfall.

Throughout the winter there is a steady succession of weather systems, three or four a week, producing cloud and plenty of precipitation. Onshore winds driven up the windward slopes of the mountains cause heavy rains along the coast. October to April is the rainy season, with most stations recording over 150 mm a month in the peak season.

By contrast, from May to September there are frequently spells of long, fine weather under the influence of a large high pressure area centred off the coast. Victoria's summers are typically dry, warm, and sunny. Indeed the Gulf Islands and the southern coast of Vancouver Island enjoy the moderate temperatures of a marine climate without the accompanying cloudy, wet weather, thanks to the protective shadow of the Olympic and Vancouver Island mountains. Residents claim it to be the most pleasant, benign climate in Canada. It is certainly the least severe.

The Interior Dryland

The British Columbia interior is located only 300 to 500 km east of the Pacific Ocean, and south of 53°N. It lies between the Coast Mountains to the west and the Rockies to the east, an area composed of a variety of plains, highlands, mountains, and picturesque lakes. The climate is continental. Winters are cold, with Arctic air spilling down from the north or flowing through the passes of the Rockies from the Prairies. Summers are warm and very dry with frequent hot days. Precipitation is distributed relatively evenly throughout the year, although brief showers and thunderstorms give an edge to the summer.

The southern interior is where we find British Columbia's, and some of Canada's, driest climates, with the most arid areas

located along valley bottoms and on east-facing slopes in the rain-shadow of the Coast Mountains. Western slopes in this region, however, are generally adequately supplied with moisture. North of the Okanagan and neighbouring valleys, winters are longer and colder and summers shorter and cooler than in the south. Precipitation is not heavy, with almost half of the yearly total falling as snow.

The North...
COMBING ARCTIC AND PACIFIC

The northern half of the province comprises two distinct regions— the interior plains and the northern plateau. The plains lie to the east of the Rocky Mountains in the basins of the Fort Nelson, Liard, and Peace rivers. They are part of the Great Basin of North America, with flat lowlands and gently rolling uplands between 900 and 1200 m. The northern plateau west of the continental divide has a more varied landscape, with mountain ranges and wide valleys. The whole area is characterized by long, cold winters and short, cool summers. Precipitation is fairly evenly distributed throughout the year, although spring tends to be the driest season and summer the wettest. Amounts are moderate, with some of the western slopes and tops of mountains receiving 400 cm or more of snowfall. Summer days are long and yearly sunshine totals ample. Summer temperatures

PRECIPITATION EXTREMES ON VANCOUVER ISLAND

Greatest number of consecutive days with precipitation	66	Cape Scott	January 197
Wettest year	8123 mm	Henderson Lake	193
Greatest number of wet days in one year	282	Holberg Fire Dept.	198
Wettest day	489 mm	Ucluelet Brynnor Mines	October 6, 196
Wettest month	2019 mm	Henderson Lake	December 192
Wettest duration in:			
5 minutes	8.2 mm	Port Alberni	May 24, 197
10 minutes	14.2 mm	Tofino	November 7, 197
15 minutes	17.8 mm	Bear Creek	December 11, 196
30 minutes	35.6 mm	Bear Creek	December 11, 196
1 hour	48.8 mm	Bear Creek	December 11, 196
2 hours	70.9 mm	Ucluelet Brynnor Mines	October 16, 196
6 hours	139.4 mm	Ucluelet Brynnor Mines	October 16, 196
12 hours	219.5 mm	Jordan River Diversion	December 25, 198
Greatest number of consecutive hours with precipitation	157	Spring Island	January 25, 197

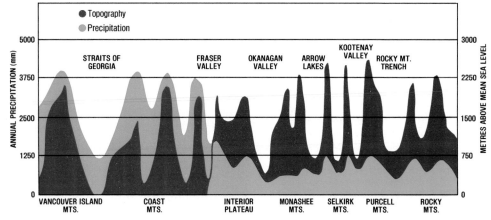

EFFECT OF TOPOGRAPHY ON PRECIPITATION ACROSS SOUTHERN BRITISH COLUMBIA (Adapted from B.C. Atlas of Resources)

- Topography
- Precipitation

STRAITS OF GEORGIA — FRASER VALLEY — OKANAGAN VALLEY — ARROW LAKES — KOOTENAY VALLEY — ROCKY MT. TRENCH

ANNUAL PRECIPITATION (mm): 5000, 3750, 2500, 1250, 0
METRES ABOVE MEAN SEA LEVEL: 3000, 2250, 1500, 750, 0

VANCOUVER ISLAND MTS. — COAST MTS. — INTERIOR PLATEAU — MONASHEE MTS. — SELKIRK MTS. — PURCELL MTS. — ROCKY MTS.

in the north are about 5°C cooler than in the south of the province. On the other hand, winter temperatures are among the lowest in the province, on average about 20°C below those in the Okanagan.

Understanding the B.C. Climate

Much of British Columbia's climate can be explained by its mid-latitude location on the east side of the Pacific Ocean and by the complex arrangement of mountains and valleys paralleling the coast.

TOPOGRAPHY

The British Columbia coast owes its moderate climate to the prevailing westerlies

blowing off the Pacific. The ocean, being warmer than the land in winter and cooler in summer, keeps temperatures within a relatively narrow but comfortable range. Rugged mountains confine this moderate, but often wet, climate to a narrow band of the province between the sea and the Coast Mountains. They also help to keep out cold waves of Arctic air that sometimes penetrate into the interior from the north and east. The coastline is indented with innumerable fiords that extend into the heart of the Coast Mountains. Mountain peaks are usually below 2400 m. On the coastal islands the mountains reach a maximum elevation of 1100 m above sea-level on the Queen Charlottes and 2100 m in the central part of Vancouver Island.

To the east, along the Alberta border, lie the majestic Rocky Mountains, with summits that reach 3000 m. This massif ekes out moisture from the Pacific maritime air, watering the west slopes but leaving the deep Rocky Mountain Trench and the eastern slopes in Alberta dry. Cold air from the north and the Prairies is kept out for most of the winter; however, on occasion the cold air surges into the province by flowing through the mountain passes.

The area between the Coast Mountains and the Rockies consists of deep valleys, high plateaux, and rugged uplands. The windward west-facing sides of the three

BOB BEAL

AVERAGE ANNUAL PRECIPITATION (mm) FOR A PORTION OF WEST END AND NORTH VANCOUVER

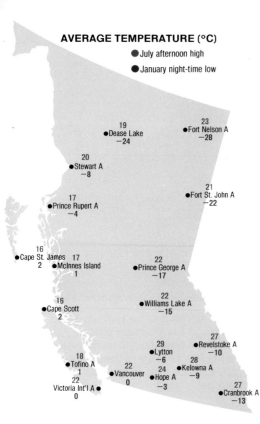

AVERAGE TEMPERATURE (°C)
- July afternoon high
- January night-time low

Dease Lake 19 −24

Fort Nelson A 23 −28

Stewart A 20 −8

Prince Rupert A 17 −4

Fort St. John A 21 −22

Cape St. James 16 2

McInnes Island 17 1

Prince George A 22 −17

Cape Scott 16 2

Williams Lake A 22 −15

Revelstoke A 27 −10

Lytton 29 −6

Kelowna A 28 −9

Tofino A 18 1

Vancouver 22 0

Hope A 24 −3

Victoria Int'l A 22 0

Cranbrook A 27 −13

intervening mountain ranges — Monashee, Selkirk and Purcell — that make up the Columbia system are moist and thickly forested. The east sides and valley bottoms are parched and covered with scrub. Overall, the amount of rainfall decreases from west to east, although snowfall shows some increase, especially over the higher Rocky Mountains.

Weather Systems

Cloudy, dull, and rainy weather prevails over the Pacific coast on most winter days, although temperatures usually remain mild. Indeed, the southwest coast of British Columbia experiences some of the warmest average January temperatures in Canada. This weather is the product of a steady stream of low pressure disturbances that originate over the Pacific, producing intense storms with gale force winds and high seas. The coast of British Columbia is a preferred track for these low pressure disturbances when they move eastward from the Pacific or southeastward from the Gulf of Alaska. In fall and winter, the frequency of such disturbances is very high. In January, for example, they influence the weather on the south coast 75% of the time.

On the remaining days, high pressure weather (clear, dry, and mild) breaks the storm monotony. Along with the cold air come brisk winds and snow. If not a taste of winter, these conditions are at least a

prelude to the next wave of Pacific storms. After striking the coast, these storms will rise over the coastal ranges, and continue eastward across the Cordillera. Inland sites are spared most of the rain and snow but experience considerable cloudiness under the presence of these Pacific air streams.

Summer, on the other hand, brings minimal storm activity along the coast. The Pacific storm track shifts farther north, taking the rain-bearing winds with it. The southern part of the coast enjoys prolonged periods of fine, warm, dry weather tempered by cool sea breezes, reminiscent of the Mediterranean. When upper atmospheric blocking persists, high pressure weather with mild Pacific air prevails, as it did in August and September during Expo '86 and during the drought of summer 1987. Occasionally a vigorous disturbance will break through, leading to a period of cloud and rain. By October, the summer weather pattern ceases, the Pacific storm track shifts southward again, and unsettled weather with nearly continuous cloud and frequent precipitation predominates.

Thermal Contrasts

Across the province the average temperature varies from -24°C in January in northeastern British Columbia to 20°C in the southern interior river valleys in July and August. The coast offers the least seasonal variation. From midsummer to midwinter, Victoria's temperature range (the difference between the highest and lowest average monthly temperatures) is only 12°C, compared with 24°C at Kelowna and 40°C at Fort Nelson.

Winter temperatures in British Columbia strongly reflect the influence of the sea and the province's varied landforms. Thanks to the ice-free Pacific, mean temperatures all along the coast remain 2 to 4°C above freezing in winter — the highest average winter temperature of any part of Canada. This is in sharp contrast to the -8°C average across the Coast Mountains and in the southern interior river valleys, and the -20°C average across the Rocky Mountains and in the northern interior. In the north, long winter nights are often very cold, and most valley stations have recorded temperatures

below -50°C — extremes not unlike those in Yukon or on the Prairies. Cold waves that penetrate the interior from the north and east stagnate in the valleys until they are dislodged by Pacific air streams. Only occasionally does Arctic air reach the coast. Vancouver has recorded a temperature of -17.8°C (0°F) — the usual mark of Canadian cold — just once, on January 14, 1950.

Across the province, temperatures are more uniform in summer than in any other season. Pacific coast summers are cooled by fresh sea breezes, so that average temperatures rarely exceed 18°C along the coast. In July the difference in average temperature between Langara, off the northern tip of the Queen Charlotte Islands, and Victoria, 1000 km to the south, is about 1°C.

Across the Coast Mountains, the temperature variability increases. Summer days are hot and nights are cool in the southern interior valleys, producing some of the greatest daily temperature ranges in Canada. Most summers will see at least one 37.8°C (100°F) reading somewhere in the B.C. interior — most likely at Lytton. In fact, the record for the highest temperature ever measured in British Columbia (and the second highest in Canada) was set in this area — 44.4°C at Lytton, Lillooet, and Chinook Cove on July 16, 1941.

In Northern British Columbia, across the Cordillera, and onto the interior plains, summer temperatures are among the lowest recorded in the province. The July mean temperature at Dease Lake is 12.5°C and at Fort Nelson, 16.6°C.

Frost

Canada's longest frost-free period, 280 days, occurs at the southern end of Vancouver Island, although 200 days is more typical of the Pacific Coast. Victoria (Gonzales Observatory) has a frost-free season that usually extends from March 1 to December 8, although there have been years without any frost at all. Moreover, at other sites in the Victoria area, but away from the ocean, there are more than 100 fewer days without frost than at Gonzales.

Across the Coast Mountains and throughout the interior valleys, the frost-free season ranges from 140 days in the south to 60 days and less in the north. In the orchards and market gardens of the Fraser, Okanagan, and neighbouring valleys, where a great variety of delicate fruits and berries are

BRITISH COLUMBIA WEATHER SUPERLATIVES

	What	Where	When
Temperature			
highest	44.4°C	Lytton	July 16, 1941
lowest	-58.9°C	Smith River, Airport	Jan. 31, 1947
Greatest ...			
precipitation	8123 mm	Henderson Lake	1931
snowfall	2447 cm	Revelstoke Mount Copeland	1971-72
number of fog days	254	Old Glory Mountain	1964
wet period	300 days	Langara	1939
number of blowing snow days	68	Old Glory Mountain	1957-58
sunshine	2426 hours	Victoria Gonzales	1970
wind (maximum hourly)	143 km/h	Bonilla Island	Feb. 20, 1974
frost-free period	685 days	Victoria Gonzales	1925-26

grown, the frost hazard is a critical concern. As a result, farmers try to reduce the threat of frost by taking advantage of local geographic features. Orchards and farms are located where air drainage, elevation, slope, water bodies, and other factors will help to moderate air temperatures. Irrigation is also used to provide needed crop moisture and to protect crops against early spring freezes. Fall frost, though not normally expected before the end of October, may also be a threat at harvest time. Especially threatening to fruit trees are the killing low temperatures that occasionally occur when outbreaks of cold Arctic air enter the province.

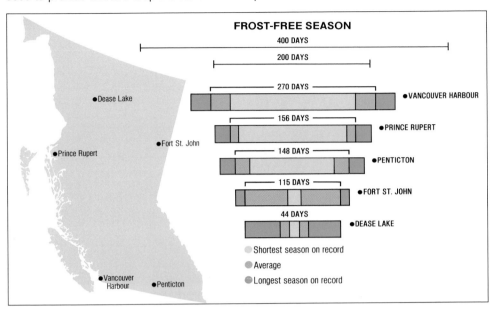

FROST-FREE SEASON

400 DAYS

200 DAYS

270 DAYS — ●VANCOUVER HARBOUR

156 DAYS — ●PRINCE RUPERT

148 DAYS — ●PENTICTON

115 DAYS — ●FORT ST. JOHN

44 DAYS — ●DEASE LAKE

Shortest season on record
Average
Longest season on record

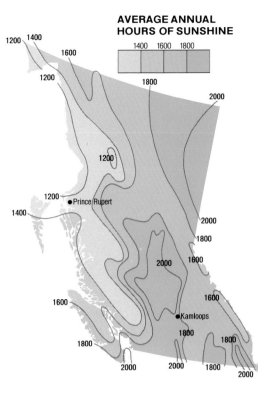

AVERAGE ANNUAL HOURS OF SUNSHINE

1400 1600 1800

●Prince Rupert

●Kamloops

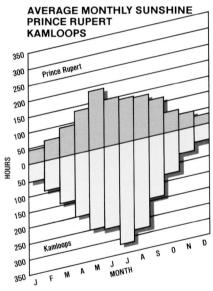

AVERAGE MONTHLY SUNSHINE PRINCE RUPERT KAMLOOPS

Prince Rupert

HOURS

Kamloops

MONTH
J F M A M J J A S O N D

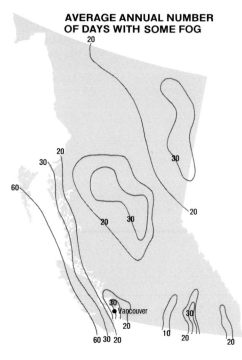

AVERAGE ANNUAL NUMBER OF DAYS WITH SOME FOG

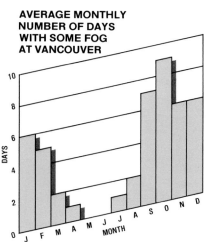

AVERAGE MONTHLY NUMBER OF DAYS WITH SOME FOG AT VANCOUVER

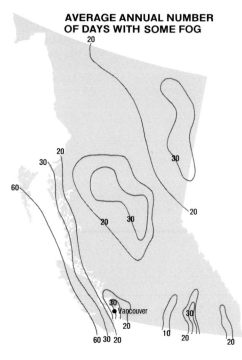

BOB CROSBY

ANNUAL NUMBER HOURS OF FOG AND SMOKE AT VANCOUVER INT. AIRPORT

Fog
Smoke

ANNUAL NUMBER OF HOURS

2500
1250
0

1947   1950   1960   1970   1980   1988

Sunny British Columbia

Sunshine is almost as variable as rainfall. Bright sunshine is highest on the lee side of the mountains but extremely low — the lowest in Canada — on the northern Pacific coast. Cranbrook, in the interior has the greatest number of hours of bright sunshine on average in the province, 2244. This is less than that recorded on the southern Prairies but above that at most locations in Ontario, Quebec, and Atlantic Canada. On the other hand, the town of Prince Rupert, on the north mainland coast, has the dubious distinction of being the least sunny of all reporting stations in Canada, with an average sunshine total of only 985 hours a year. This is only 27% of the total sunshine possible at this location. Most of the time, the sun is blocked by the overcast skies,

rain, and fog that accompany the frequent Pacific storms that pass over the town.

Victoria's location in the rain shadow of the Olympic Mountains gives it more sunshine than any other coastal location, about 2100 hours on average, adding to its favour among Canadians as a resort and retirement city. Vancouver is near the Canadian average of 1900 hours a year.

Foggy British Columbia

Exposed localities along the outer Pacific coast experience more than 60 days of fog

each year, frequent enough to cause serious transportation delays but not nearly as disruptive as in the Atlantic region, where 120 to 180 fog days occur each year. Fog occurs when moist, stable air moves across the cooler sea waters. Although this can happen at any time during the year, it is most prevalent in summer and fall.

Again, topography is a strong control. Fog enters certain deep inlets but is shunted away from islands and protected inlets along the east side of major islands. For example, Cape St. James at the southern tip of the

Penticton rarely has fog, but to the north, in the central interior, dense fog is notorious. Quesnel receives the most — 75 days a year on average and at least 10 days a month in August, September, and October. Valley fog is common in northern British Columbia, occurring most often in late summer and fall. During the winter, far northern stations report the occasional occurrence of ice fog on very cold days.

Changing Winds

The Queen Charlotte Islands is one of the windiest regions in Canada. Gales blow an average of 120 days a year at Cape St. James — a national high. The average wind speed there is 33.7 km/h (41.7 km/h in December and 26.1 km/h in July). This is considerably higher than in the protected interior valleys, where Penticton, for example, averages 12.7 km/h and Castlegar, 8.3 km/h. At Cape St. James and elsewhere along the open coast, calms occur rarely — about 1% of the time in the summer and less often in winter. Sea breezes are a notable feature of the coastal summer day.

Topography exerts a strong influence on wind direction. Winter winds along the coast are overwhelmingly from the east and southeast. At Vancouver, easterlies account for 52% of all winds in winter and 44% of the winds year-round. At most other stations along the inner and outer coast, the prevailing winds become westerly with the arrival of summer. In the sheltered interior valleys, the wind usually blows along the length of the valley and thus generally in a north or south direction. Calms are frequent here, especially during winter nights.

Severe Weather

YES AND NO!

Violent storms are not rare in British Columbia; however, thunderstorms and tornadoes lack the intensity and frequency of such storms east of the Rockies.

Hurricanes are almost non-existent because the cool waters of the North Pacific usually prevent the hot, humid air of the tropics from reaching British Columbia. One of the few to get through was typhoon Freda, or rather its remnants, which struck the Pacific Northwest on October 12, 1962. At Victoria, winds reached sustained speeds of 74 km/h with gusts to 145 km/h. Damage estimates were $10 million ($48 million in 1990 dollars) to store windows, small boats, hydro-lines, and gardens. Seven deaths were attributed to the storm.

Thunder is about as unheard along the Pacific coast as it is in Atlantic Canada. Victoria averages two thunderstorms a year and Vancouver about six. Many years pass with none at all. Thunder is much more

Queen Charlottes has 110 fog days a year, whereas Sandspit on the east coast of the Queen Charlottes has only 20. Some of the more exposed sites include Tofino with 93 fog days, Cape Scott with 88, and Langara with 45. Comox and Nanaimo with only 20 fog days each and Port Hardy with 34 are typical of the sheltered locations.

Vancouver and the Fraser Delta have a high incidence of fog, which occurs 14% of the time at these locations. Fog is even more of a problem in winter, occurring for one quarter of the time and often persisting throughout the day and for long spells. One of the most annoying foggy spells on record occurred during the holiday period in late December 1985, when thick, persistent fog severely hampered airport operations and forced travellers to change their Christmas holiday plans.

Fog is not just a problem along the coast. Inland, fog days may number 60 or more a year in the sheltered valleys, where dense fog occurs most often at night and in the early morning in the fall and winter. Castlegar in the southeast has 46 days a year with fog — at least 5 days in each of the months from September to March. To the west,

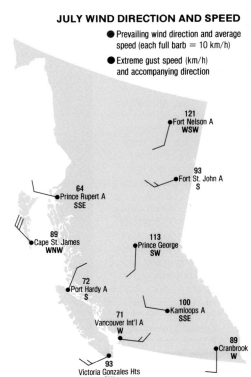

JANUARY WIND DIRECTION AND SPEED

- Prevailing wind direction and average speed (each full barb = 10 km/h)
- Extreme gust speed (km/h) and accompanying direction

74 Fort Nelson A NW

138 Fort St. John A SW

109 Prince Rupert A WSW

137 Cape St. James SE

129 Prince George S

113 Port Hardy A ESE

113 Kamloops A N

97 Vancouver Int'l A SW

64 Cranbrook A WNW

129 Victoria Gonzales Hts SW

JULY WIND DIRECTION AND SPEED

- Prevailing wind direction and average speed (each full barb = 10 km/h)
- Extreme gust speed (km/h) and accompanying direction

121 Fort Nelson A WSW

93 Fort St. John A S

64 Prince Rupert A SSE

89 Cape St. James WNW

113 Prince George SW

72 Port Hardy A S

100 Kamloops A SSE

71 Vancouver Int'l A W

89 Cranbrook W

93 Victoria Gonzales Hts SW

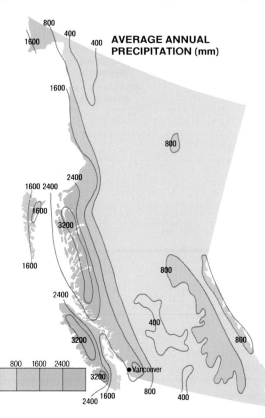

AVERAGE ANNUAL PRECIPITATION (mm)

there, with July being the most likely month for an occurrence. Vancouver has had more than its share of tornado touchdowns — in 1956, 1962, and 1985.

Fall is the wettest and stormiest season, especially along the north coast. Wind and rain associated with intense mid-latitude disturbances can be destructive and life-threatening. In fall and winter new storms moving across the Pacific reach the coast about every two days, bringing renewals of overhanging cloud, abundant rain, and buffeting winds. The procession breaks down rapidly after February and storm activity reaches a minimum in June.

Heavy rains and snow melt have caused considerable flooding in the Fraser Valley and delta. The heavy rains can also trigger landslides, mudslides, and avalanches, causing further damage and loss of life.

Wet and Dry

B.C. HAS IT ALL!

British Columbia is renowned for its precipitation contrasts. Some places on Vancouver Island have the heaviest rainfalls in Canada, with annual totals of more than 2500 mm. Apart from short-duration rainfall, every

record for greatest rainfall is held by ocean-facing stations on Vancouver Island. However, the image of "rain all the time" on the West Coast is wrong. Although precipitation is indeed heavy in the winter, summers along the coast are markedly dry. At Vancouver only 10% of the annual total precipitation falls during the summer and dry spells are frequent. Dry days outnumber wet days 5:1 in July and August.

Rain becomes heavier along the coast in September and peaks from October to December, when monthly totals generally exceed 250 mm and may exceed 400 mm locally. Nearly all the rain comes in prolonged periods of moderate intensity from a succession of low pressure disturbances or frontal systems moving off the Pacific. In the fall and winter, wet days are twice as frequent as dry ones, which, if not wet, tend to be overcast at least. At the high elevations, most of the winter precipitation falls as snow, encouraging winter recreation in accessible areas.

The heaviest precipitation occurs on the outer coast of Vancouver Island, the Queen Charlotte Islands, and on that part of the mainland that lies north of Vancouver Island.

frequent in the interior along the upper reaches of the Fraser River, with 20 thunderstorm days a year. Most storms develop in the summer afternoon with the heating of the land. With its cooler summer, the north has less thunder and lightning, but still enough to present a forest fire risk.

Hail is common enough to pose an occasional threat to orchard and garden farming in the south, but tornadoes are seldom seen. A study of 30 years (1950-79) of tornado sightings in Canada found 20 in British Columbia, but none were of a severe nature and there were no known injuries or deaths. The southern interior has the highest tornado risk in the province. About 1 tornado every 2 years per 10 000 km can be expected

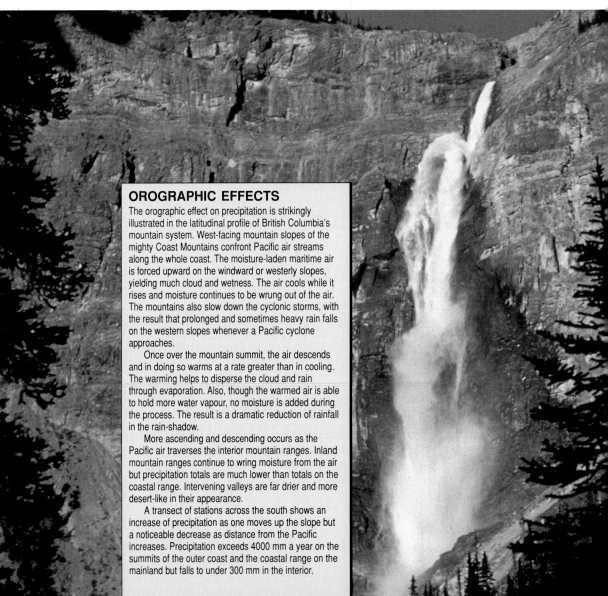

OROGRAPHIC EFFECTS

The orographic effect on precipitation is strikingly illustrated in the latitudinal profile of British Columbia's mountain system. West-facing mountain slopes of the mighty Coast Mountains confront Pacific air streams along the whole coast. The moisture-laden maritime air is forced upward on the windward or westerly slopes, yielding much cloud and wetness. The air cools while it rises and moisture continues to be wrung out of the air. The mountains also slow down the cyclonic storms, with the result that prolonged and sometimes heavy rain falls on the western slopes whenever a Pacific cyclone approaches.

Once over the mountain summit, the air descends and in doing so warms at a rate greater than in cooling. The warming helps to disperse the cloud and rain through evaporation. Also, though the warmed air is able to hold more water vapour, no moisture is added during the process. The result is a dramatic reduction of rainfall in the rain-shadow.

More ascending and descending occurs as the Pacific air traverses the interior mountain ranges. Inland mountain ranges continue to wring moisture from the air but precipitation totals are much lower than totals on the coastal range. Intervening valleys are far drier and more desert-like in their appearance.

A transect of stations across the south shows an increase of precipitation as one moves up the slope but a noticeable decrease as distance from the Pacific increases. Precipitation exceeds 4000 mm a year on the summits of the outer coast and the coastal range on the mainland but falls to under 300 mm in the interior.

In these areas the average yearly precipitation may be in excess of 2500 mm. But the West Coast is not without its dry climates. In the lee of the Olympic Range, there is a remarkable rain-shadow over the Saanich Peninsula of Vancouver Island. Victoria's driest station receives only 635 mm of precipitation a year, in comparison with 1113 mm at Vancouver International Airport just across the Strait of Georgia.

The southern interior provides a striking contrast with the Pacific coast in almost every aspect: amount, seasonality, variability, form, and origin.

As we have already seen, scanty precipitation is the main feature of the southern interior valleys and leeward slopes. In the bottoms of the Fraser, Thompson, Okanagan, and neighbouring valleys, and in the Rocky Mountain Trench, total precipitation is less than 250 mm a year. Kamloops, Ashcroft, and Penticton are some of the driest stations, with amounts between 200 and 300 mm a year. The western slopes of the interior mountains are wetter, with annual amounts exceeding 800 mm.

The higher elevations and more exposed locations get their moisture chiefly

from passing maritime air. Consequently, by far the greatest part of their precipitation comes in the winter when numerous Pacific storms hit the province. Along the valley floors, precipitation is greater in summer.

Variable precipitation is typical of most dry climates, and the southern interior of British Columbia is no exception. Nevertheless, some of Canada's most productive orchards flourish here, but irrigation is required to offset severe moisture shortages.

In the interior a much greater percentage of the annual precipitation (about 40 to 50%) falls as snow than on the coast. Summer rains in the interior come mostly from afternoon thunderstorms rather than from Pacific disturbances and fronts.

Farther north in the interior, rain and snow amounts continue to reflect the relief, with windward slopes being wet and leeward slopes being dry. Yearly amounts are much below those on the coast. Precipitation is fairly evenly distributed throughout the year, though spring is drier than the other seasons and summer and fall are somewhat wetter. Precipitation totals increase from 327 mm at Atlin to 406 mm at Dease Lake. Over the northeast plains, yearly totals are 450 mm and greater.

Big Snowfalls

Snow is exceedingly important to the economy of British Columbia. Ample snowfall and prolonged snowcover mean a profitable ski season in the mountain resorts north of Vancouver and guarantee waters for hydro-electricity, irrigation, and forestry. The greatest snowfalls occur on the higher elevations of the west-facing slopes of the major mountain ranges, particularly along the B.C. – Alaska Panhandle boundary and in the Cariboo – Monashee – Selkirk – Purcell group. Along the coast only 5% of the year's precipitation falls as snow, whereas in the valleys of the Columbia, Caribou, and Rocky systems colder temperatures mean a greater percentage of the year's precipitation (40%) occurs as snow.

Pacific disturbances bring the moisture for the snow. In the Cariboos between 750 and 1000 cm of snow will fall in most winters. Avalanches pose a threat, as do spring and summer melts that produce roaring torrents. At valley bottoms, snowfall amounts to 35% of annual precipitation but the actual amount of snow is only about 100 cm. At Glacier, in the north Columbia Basin, (altitude 1875 m) the seasonal snowfall averages 1433 cm, the most for any station in Canada.

Along the coast, snowfall is light in the lowland areas and does not last long.

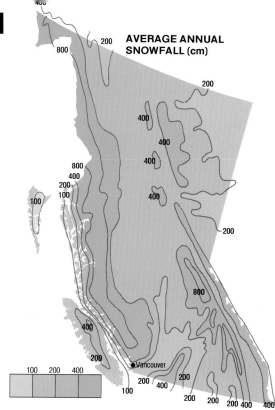

AVERAGE ANNUAL SNOWFALL (cm)

Temperatures are too mild for much snow to fall or stay near sea-level. However, since small changes in temperature can make the difference between rain and snow, the monthly snowfall amounts are quite variable. Snowfall increases inland from the coast and with increasing elevation. Snowfall accounts for just over 3% of the average annual precipitation at Cape St. James and for about 5% at Vancouver and Victoria.

In the northern interior the winters are considerably colder than in the south. The snow season lasts two to three months longer, but snowfall totals are generally less than at comparable elevations in the south. Snow in the valley bottoms averages 125 to 150 cm and on the west slopes of the uplands, at least three times as much (400 cm).

Despite the scanty snowfalls in British Columbia's major cities, when snow does occur, it makes headlines — especially in eastern Canada. One of Vancouver's worst storms occurred on January 19, 1935. More than 40 cm of snow fell on the city. Strong winds whipped the snow into 2-m drifts before copious rains — 267 mm over 4 days — washed them away. Victoria experienced its snowiest winter in 1915-16 when 196 cm of snow fell. On February 13, 1916 the city received its greatest one-day snowfall ever, 55.3 cm. Victoria's only snowplough was kept busy amidst abandoned cars and stalled transit vehicles. The fire department had to use horse-drawn wagons.

Canada's Cold Pole

Yukon's spectacular scenic beauty owes much to the territory's climate. Snow-capped peaks, majestic ice-fields, mighty waterways, and a profusion of bright blooms and delicately tinted skies are largely the product of climate.

Yukon winters are white and long. Snow blankets the ground and ice covers the rivers for seven months. Days are short, but skies are bright, calm, clear, and dry.

Summer is colourful though brief. Snow and ice are gone by the first of June, and the long days promote rapid plant growth. The skies continue bright, calm, clear, and dry.

Yukon's climate is generally continental. The Beaufort Sea is ice-covered too long to moderate air temperatures, and the Pacific

Ocean is cut off to the south by the rugged St. Elias Mountains. However, frequent wintertime incursions of mild Pacific air do spill over and offer a welcome respite from the cold Arctic air.

The most noteworthy facts about the Yukon climate are the enormous annual temperature range and the remarkably high air pressure, both of which are greater than anywhere else in North America. The average annual temperature range (the difference between the average temperatures of the warmest and coldest months) is around 40°C. The greatest absolute range (the difference between extreme high and extreme low) yet recorded is 98.3°C at Mayo. In contrast, the average and absolute range of temperature at Vancouver is 20° and 51°C, respectively. Dawson holds the Canadian record for highest sea-level pressure, 107.96 KPa, set on February 2, 1989. Normal January sea-level pressure is 102.31 kPa.

Topography has a strong influence on Yukon's climate. Throughout the territory, mountains have more effect on the climate than any other controls, including latitude, marine effects, and weather systems. Mountain ranges can channel winds, block storms, prolong weather spells, enhance precipitation on windward slopes, or suppress precipitation on downwind slopes. The most imposing mountains are the St. Elias and Coast Mountains, enormous barriers to Pacific storms and mildness. They include Mount Logan, which at 5951 m is Canada's highest mountain. On the east side, the Mackenzie-Richardson mountains are less formidable barriers to the air masses from the Arctic. Between these two mountain blocks lies the Interior Basin — an irregular, rough highland ranging in elevation from 900 to 1500 m and interrupted frequently by local mountains, deep, narrow valleys, rugged plateaux, and long, deep, narrow, glacier-fed lakes.

Yukon Weather

In winter, Yukon is generally cold and dry, owing to the dominance of cold Arctic air that stalls over the central interior for weeks on end. The cold air moves in from Alaska and the Northwest Territories, occupying the interior basin and deep valleys where it

becomes even colder as a result of heat loss by radiation during the long northern nights. Often portions of the Arctic air mass will move southward, causing cold outbreaks across most of western Canada.

During the winter months, a nearly permanent low pressure centre in the northern Gulf of Alaska steers a succession of storms into the southern half of Yukon. In spring the storm track shifts northward and the intensity of the disturbances diminishes.

In summer, Yukon's weather is influenced by a large high pressure cell located over the mid-Pacific and by a weak low pressure system over the northern Bering Sea, west of Alaska. The storm track, although weakened, creates significant cloud and precipitation, especially in the north. Although there is considerable summer sunshine, June-to-August rainfall usually exceeds total precipitation during the six months of winter, even though the number of wet days is identical in summer and winter.

During the fall, the contrast between Arctic air and maritime Pacific air increases and storms intensify. The major storm tracks gradually move southward as the fall progresses, with considerable cloud and precipitation occurring in southeastern Yukon. By mid-December, after general freeze-up, Arctic highs dominate and the showery weather ends.

Yukon Days and Yukon Nights

Yukon extends over 10 degrees of latitude from 60 to nearly 70°N. Latitude is an important climate determinant because of its direct link to the length of day and the amount of solar radiation.

In winter in the south, sunlight lasts only a few hours a day, but the north, situated above the Arctic Circle, gets no direct sunlight at all. At Whitehorse the least amount of daylight possible is less than 6 hours on December 21. On the Arctic coast, polar night lasts for 7 weeks from November 29 to January 15. Polar night is the period, north of the Arctic Circle (66°33'N), when the sun does not rise above the horizon. It is seldom pitch black, however. Nights are illuminated by a surprising amount of sky light, moonlight, and the northern lights (the aurora borealis).

In April, day length increases by more than 10 minutes a day. Summer is a sunlit world in Yukon. On the Arctic Coast, the sun is up continuously from about May 19 to July 26, and at Whitehorse the longest day extends over 19 hours, from about 3:30 a.m. to 10:30 p.m.

The increased length of summer days helps to compensate for the decreased intensity of solar radiation due to the relatively low angle at which the sun's rays strike the earth at higher latitudes. As a result, the total incoming solar energy available during the summer is approximately equal to that at temperate latitudes. The prolonged light exposure promotes rapid plant growth for such crops as spinach, winter barley and wheat, oats, grasses, and clover. The longer days also allow more heat to accumulate, producing warmer temperatures than might be expected at higher latitudes.

Ice Fog

Fog in Yukon occurs principally from late summer to early winter when there is still enough open water to supply moisture to the dry Arctic air.

Adding to the late fall/early winter peak in fog frequency is the formation of ice-crystal fog, especially near townsites and airports. This is formed in frigid air below -30°C, when the air can hold very little water vapour. When cooled further, the minute amounts of supercooled water vapour condense into a fog of ice crystals rather than water droplets. Ice fog commonly occurs where there is a source (usually artificial) of extra moisture. Artificial moisture sources include the release of water in aircraft exhaust, moisture by-products from other types of fuel combustion, and vapour expelled through breathing. In settlements, periods of ice fog occur during the coldest spells, when the amount of fuel consumed for space heating increases, and persist the longest during calm or light winds and in sheltered valleys. Historical records point to an upward trend in ice fog frequency while industrial activity expands and air traffic increases.

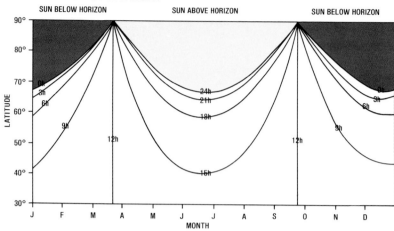

DURATION OF DAYLIGHT

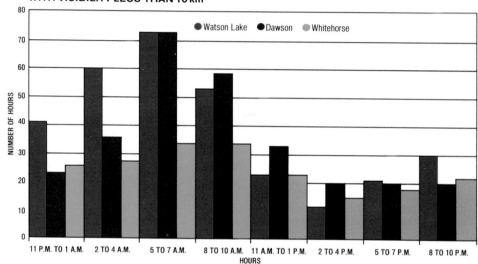

ANNUAL NUMBER OF HOURLY FOG OCCURRENCES WITH VISIBILITY LESS THAN 10 km

● Watson Lake ● Dawson ● Whitehorse

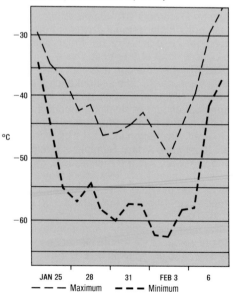

TEMPERATURE DURING THE RECORD-BREAKING COLD SPELL AT SNAG (FEBRUARY 3, 1947)

- - - Maximum   - - Minimum

METEOROLOGICAL MOMENT

At 7:20 a.m. Yukon time, the observer at the Snag weather station scratched the side of the thermometer sheath to mark the low temperature point that day, February 3, 1947. The reading was -63°C. When the thermometer was recalibrated three months later, the temperature was confirmed. It was an all-time record low for Canada and for North America. At the time, skies were clear, the winds were calm, and the visibility was unlimited. The previous morning's record of -62°C had stood for but a day.

During the cold spell, weather staff recalled that audibility was increased, there was intense radio static, fog patches hovered above the dog teams, the freezing of one's breath made a hissing sound, and vapour trails were suspended at 100 to 500 m for 3 or 4 minutes.

Lowest but not the Coldest

In comparison with the eastern Arctic, winters are remarkably mild in Yukon, but when the temperature drops, it usually falls lower than anywhere else in Canada. In fact, the lowest temperature ever recorded in North America, -63°C, was reported at Snag, Yukon, on February 3, 1947.

At times Pacific air may edge into the southwest of the territory, resulting in intervals of mild weather which may last for days or weeks. January is the coldest month, except on the narrow Arctic slope along the Beaufort Sea coast where the coldest weather occurs in February. Average January temperatures range from -

A. J. BRAZEL

20°C in the south to -30°C in the north, with lower values in the interior valleys.

An interesting feature of a Yukon winter is the almost permanent reversal of the normal pattern in which temperature decreases with elevation. Surface temperatures above ice- and snow-covered surfaces drop very rapidly, especially during prolonged periods of clear skies and calm or light winds. Air in contact with the cold surface loses heat to the atmosphere through radiation cooling. As a result, a shallow, intense inversion is present almost continuously in Yukon from late October to early March and on occasion in the summer. The increase of temperature with height in these inversions can be confined to a few tens of metres or be in excess of thousands of metres. Under extreme conditions, valley temperatures of -45°C can rise through the inversion to -15°C at the 1000-m level and to near 0°C at 1500 to 2000 m.

Owing to the temperature inversion, higher elevations often experience less severe winter temperatures and more frequent midwinter mild spells than places situated in valleys. For example, in only six

years. Faro (elevation 691 m) has recorded temperatures as low as -55°C, wheras in over 13 years Anvil (elevation 1158 m) has recorded nothing lower than -46.1°C.

Fortunately, spells of intense cold are usually of short duration, especially in the south. Over a series of 30 Januarys at Whitehorse, there were 60 separate occasions when the temperature dropped below -40°C. Only one occurrence lasted as long as 10 days. Farther north at Mayo such cold spells occurred on 137 occasions, of which 3 lasted 15 days or more, and the longest was 24 days. Generally, most severe cold spells are accompanied by calm or light winds, reducing the likelihood of severe wind chill.

The transition period from winter to summer and vice versa is remarkably short. Most people say May is spring and September is fall. Although brief, summers are pleasant but they may be changeable. Except in the far north and in areas of elevated terrain, all stations have average temperatures above 10°C during June, July, and August. Further, each Yukon summer day can have as much as four hours more sunshine than the same days in southern Canada. Summer temperatures are only slightly cooler than at many localities in western Canada. For example, the average July temperature at Dawson is 15.6°C, whereas at Edmonton it is 15.8°C. Winter temperatures in the north are much colder. Dawson's mean January temperature is -30.7°C and Whitehorse's is -20.7°C. Edmonton's -16.5°C seems almost mild in comparison

The average annual temperature is below zero in all areas and generally decreases as latitude increases. Freezing temperatures have occurred in every month at all Yukon stations except Watson Lake.

The frost-free season is short and highly variable. This season ranges from an average of 93 days at Watson Lake to only 19 days at Swift River; both sites are at nearly the same latitude and elevation. Haines Junction has one of the shortest frost-free seasons, only 21 days, yet one year it had 63 consecutive days without frost.

Along some major river valleys, where the warmest summer weather is found, a viable agriculture of cool-season vegetables and forage crops is practised. Dawson has more than 1000 growing degree-days, the highest in the Territory.

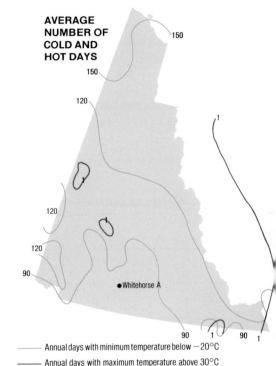

AVERAGE NUMBER OF COLD AND HOT DAYS

—— Annual days with minimum temperature below −20°C

—— Annual days with maximum temperature above 30°C

AVERAGE FROST-FREE PERIOD (DAYS)

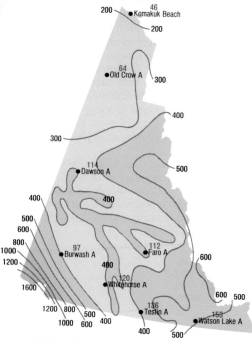

AVERAGE ANNUAL PRECIPITATION (mm)

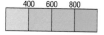

● Average annual number of wet days

AVERAGE MONTHLY SNOWFALL (cm)

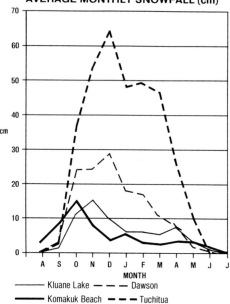

——— Kluane Lake   — — — Dawson

▬▬▬ Komakuk Beach   - - - - Tuchitua

Wet and Dry All Over

Precipitation throughout Yukon varies considerably from north to south and from valley bottom to mountain peak. Annual amounts are light over the interior, ranging from 200 mm along the Arctic coast to 700 mm on the south slopes of the Mackenzie Mountains. The southwest corner of the Territory is exposed to storms from the Pacific and receives about 2000 mm of precipitation annually. Orographic effects are very evident. Generally speaking, the western and southern slopes of the major mountains receive the greatest amounts of precipitation, wheras the precipitation-shadows to the north and east of the major ranges are drier.

A representative territory-wide precipitation total would be 330 mm, which is what Teslin normally gets each year. It is also the average amount for the dry southeast corner of Alberta.

Generally, at most stations, July and August are the wettest months and April is the driest. In the south, the wet period extends into October and November, mainly owing to low pressure disturbances off the Gulf of Alaska. Summer rainfall is more abundant in Yukon than in other parts of Canada's northland. Normal monthly amounts for June, July, and August exceed 25 mm and, at a few stations, 50 mm, which is 15 to 30 mm more than at stations over the Arctic Islands. Light summer rains and overcast weather accompany frequent Pacific lows that move inland across Yukon. About 60% of the yearly precipitation at lowland sites is rain and drizzle, the remainder snow. At elevations above 1000 m liquid and solid precipitation is evenly divided, and above 2500 m precipitation is almost exclusively snow.

Winter snowfall averages less than 100 cm in the north along the Arctic slope and is generally light over the interior plateaux and valleys. Komakuk Beach has the least snowfall of any station, 60 cm, but has the longest snow season. Heavier amounts, between 150 and 250 cm, occur in the Liard Basin, in the St. Elias Mountains, over the higher Cassiar Mountains, and on the western slopes of the Mackenzie Mountains.

May is the month when the snow melts in the lower valleys — early May in the south, mid-May in central Yukon, and late May in the north.

Arctic Boomers

Thunderstorms are more numerous than might be expected in high latitudes. Many thunderstorms go unnoticed in this lightly populated region, but it is known that the frequency of thunderstorms increases with elevation. Thunder is rarely if ever heard on the north coast. The greatest number of thunder days, 11 a year, occurs at Watson Lake, wheras most stations in the interior report an average of 4 to 6 a year. The season is from late May to early September, and the prime hours are between 3 and 6 p.m. The intensity of thunderstorms doesn't compare with that in southern Canada; nonetheless, the accompanying lightning flashes are capable of setting off

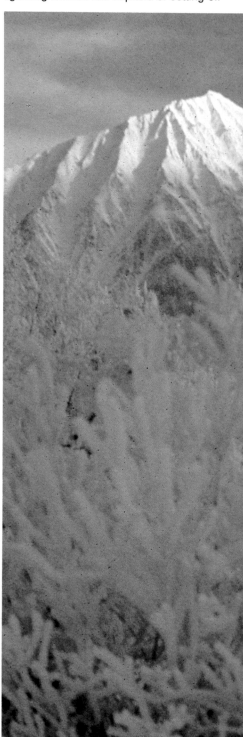

LOUISE BRENEMAN

enormous wild fires that consume large tracts of forest land.

Midnight Sunshine

Sunshine totals are not unlike those found in the Maritime provinces, although the seasonal distribution is different in the two regions. Mountain terrain and open lakes and rivers promote cloud activity in the summer and fall seasons. Cloud amounts diminish between December and April, when a high pressure circulation dominates. However, because of their longer days, the sunniest

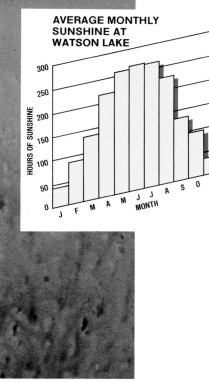

AVERAGE GROWING DEGREE-DAYS

	Accumulated average daily temperature above 5°C
Whitehorse	**897**
Dawson	**1015**
Komakuk Beach	**187**
Okanagan	2040
Edmonton	1560
Winnipeg	1785
Windsor	2533
St. John's	1196

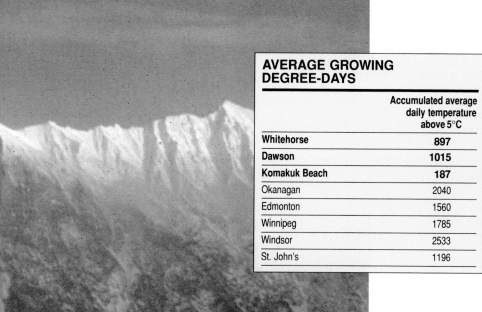

AVERAGE MONTHLY SUNSHINE AT WATSON LAKE

HOURS OF SUNSHINE

300
250
200
150
100
50
0

J F M A M J J A S O N D
MONTH

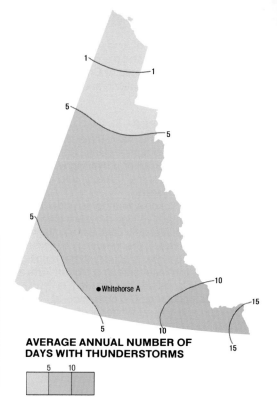

AVERAGE ANNUAL NUMBER OF DAYS WITH THUNDERSTORMS

5   10

AVERAGE ANNUAL HOURS OF SUNSHINE

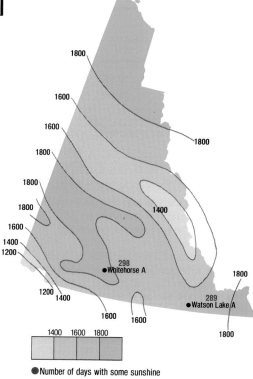

1400   1600   1800

● Number of days with some sunshine

months are May and June, with an average of 280 hours of sunshine a month. Least sunshine is experienced in December, when the monthly total duration varies from 0 to 30 hours.

Yukon Breezes

Yukon wind patterns have three characteristics: a high proportion of calms, an absence of any one preferred or prevailing wind direction throughout the Territory, and a low average wind speed.

In winter, Yukon's weather is under the control of a massive high pressure area, with its cushion of intensely cold, dense, stagnating air. As a result, the frequency of calms, both day and night, is usually large, ranging from 19% at Whitehorse to a high of 79% at Dawson. Calms are less frequent in summer, occurring 50% of the time at Dawson and 10% at Whitehorse. Summer calms also happen more often at night.

Wind measuring stations, located principally in sheltered valleys, are not well sited for detecting the regional air movement. Often, prevailing directions from

comes from melting snows on the highlands in early summer and melting ice in the glacier and ice-fields in late summer. Since most of Yukon's limited precipitation also falls during the summer, river levels are highest between May and October. June is generally the month with the highest runoff. However, the sudden rise of water levels in May, when streams are clogged with ice, can cause serious flooding. Such was the situation on May 3, 1979, when an ice jam on the Yukon River caused severe flooding in Dawson. Waters rose 2 m, submerging several historic sites and damaging 80% of the town's buildings.

Summer rainfall is not large and is fairly evenly distributed throughout the Territory. Heavy rainfalls occur occasionally, however, and can lead to disastrous flooding, such as occurred on July 15-16, 1974, when sections of the Alaska Highway and many bridges were washed away in the worst

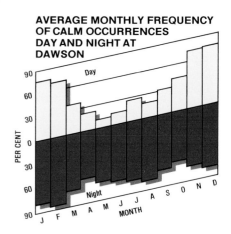

AVERAGE MONTHLY FREQUENCY OF CALM OCCURRENCES DAY AND NIGHT AT DAWSON

storm in the highway's history. During the storm, the area around Watson Lake received 100 mm of rain, with as much as 25 mm falling in a six-hour period.

YUKON WEATHER SUPERLATIVES

	What	Where	When
Temperature			
highest	36.1°C	Mayo	June 14, 1969
lowest	-63.0°C	Snag	Feb. 3, 1947
Greatest ...			
precipitation	677 mm	Tuchitua	1968
snowfall	452 cm	Tuchitua	1971-72
number of fog days	86	Komakuk Beach	1976
wet period	186 days	Watson Lake	1966
number of blowing snow days	107	Komakuk Beach	1982-83
sunshine	2064 hours	Haines Junction	1972
wind (maximum hourly)	108 km/h	Shingle Point	Dec. 31, 1973
frost-free period	140 days	Carcross	1946

these sites coincide with the orientation of valleys and change little year-round. Consequently, a variety of prevailing directions occurs.

North coastal sites have the highest average wind speed and the greatest wind extremes. On the north slope at Komakuk Beach, winds above 30 km/h occur 20% of the time. Inland at Mayo, Snag, and Teslin, only 1% of the winds exceed 30 km/h.

In winter, when there is a high frequency of inversions and light winds, air pollution can last for several days. The central Yukon basin, where ventilation is low, is especially prone to lengthy pollution episodes in winter.

Spring Surpluses

Even though rainfall is sparse in most of Yukon, its rivers are still supplied with generous amounts of water. This water

LOIS TREMBLAY

Canada's Frozen Desert

ENVIRONMENT CANADA

The Canadian northland spreads over half a continent, covering more than one third of the total area of Canada. Its land area would easily encompass western Europe, British Columbia, Ontario, and Quebec combined. Arctic waters claimed by Canada cover 20% of the total area of the Arctic Ocean. The coastline perimeter is longer than that of all the ocean-fronting provinces together. With such a vast area, it should come as no surprise that the diversity in scenery is remarkable: taiga forests and treeless tundra, lush meadows and rocky outcrops, active permafrost and permanent ice-caps, archipelago lowlands and fiorded coastlines.

What is surprising, given the size of the Territories, is the lack of climate diversity. The climate is harsh everywhere. Long, dark, and cold winters are interrupted by short, cool, and cloudy summers. Precipitation is light, totalling a territory-wide average of 100 mm, and is usually concentrated in the warmer months. There are exceptions to this climate generalization: the southern Mackenzie has summer average temperatures not unlike those in the Prairies, and the eastern coast of Baffin Island has 400 to 600 mm of precipitation in an average year.

Despite the generally scanty precipitation, moisture is plentiful — in lakes and rivers, in the muskeg and permafrost, in the snowcover, and in the permanent ice or in the Arctic Sea. There is a growing season, but it is short. Temperatures are generally below -18°C for five months or more, and the annual range of temperatures is one of the greatest in North America. Winter nights and summer days are long, reaching a maximum of 24 hours.

Although lacking diversity, the North has a distinctive climate that is manifested in numerous environmental phenomena: permafrost, the polar ice-pack, the Arctic desert, the Arctic inversion, Arctic haze, and polar night.

Making The Climate

The climate of the Northwest Territories reflects the usual broad controlling factors: latitude and solar energy, weather systems, topography, and the nature of sea and land surfaces.

HIGH NORTH

Latitude is the dominant influence on the climate of the Northland. Around Great Slave Lake, the amount of possible sunlight varies from 5 hours at the start of winter to 20 hours at the start of summer. North of the Arctic Circle, at Coppermine, the sun does not set in midsummer nor rise in

---

METEOROLOGICAL MOMENT

Inclement weather kept Pope John Paul II from visiting Fort Simpson, Northwest Territories, on September 18, 1984. About 3000 people, most of them native Indians, many of them travelling hundreds of kilometres by boat, plane, and vehicle, converged on this northern town for the day. But because of thick fog, the Pope's plane could not land. For 12 hours during the day visibility was near zero. Before the fog cleared, His Holiness had to move on. A disappointed Pontiff vowed to return for a short visit during his United States tour in 1987. As he had promised, three years later on September 20, 1987 the Pope flew into Fort Simpson and met with peoples of the Northwest Territories. The day began with a cold rain, but a brilliant rainbow by midmorning signalled a change to sunshine for most of the day.

midwinter. The contrast between the winter and summer versions of midnight is not normally as great as southerners expect it to be. In summer, midnight sun shines more as twilight, and in winter, the moon's reflection on the ice and snow illuminates the landscape from horizon to horizon.

At high latitudes the sun's energy arrives at such a low angle that the warmth received on a flat surface is fairly small. In summer this limitation is balanced by the increased length of day. Consequently, the total energy available in June and July is approximately the same as at temperate latitudes. The success of forage crops in the Slave, Liard, and Mackenzie River valleys is due in large part to the longer summer days, which promote rapid growth.

In spring as much as 85% of the solar energy striking the snow and ice cover is reflected directly back to space. An appreciable fraction of the remainder is used in the melting and evaporation of snow and ice. In summer, extensive cloud cover and ice-filled seas reflect more than 50% of the sun's radiation. In winter the snow surface radiates more energy back to space, resulting in steady net losses over the many winter months and contributing to the formation of highly stable, cold air masses. The net result is a long, cold winter season and a short, cool summer season.

NORTHLAND

Major topographic barriers bracket the Territories on the west and east. To the west of the Mackenzie River, the western Cordillera drops from a height of 1500 m to 150 m within a distance of 100 km, effectively making the Mackenzie Basin a rain-shadow. In the east, the rugged mountains on Baffin, Devon, and Ellesmere islands, with peaks rising to 2800 m, cause considerably increased precipitation on the eastern slopes of the islands, and block the weather systems that enter the North through Davis Strait.

Across the Archipelago and over the mainland, elevations are mostly below 500 m. As a result, topography fashions local climate but has limited effect on regional weather.

Cloudiness, precipitation, and wind speeds generally increase with elevation, whereas air temperature and moisture content decrease. However, under cooling conditions, colder air tends to flow into and accumulate in valleys, causing strong inversions that favour the accumulation of pollutants and the formation of ice fog.

WEATHER SYSTEMS

The climate of the Northwest Territories is closely influenced by seasonal changes in the general circulation of the atmosphere. Precipitation, for example, shows marked seasonal contrasts. The driest month is February, when high pressure dominates the North and low pressure centres generally cross Canada at latitudes well south of the Territories, only affecting the North while they recurve into the Davis Strait — Baffin Bay region. August is the wettest month almost everywhere in the North. At this time, storms move directly across the Territories from the west or southwest, bringing warm, moisture-laden air with them.

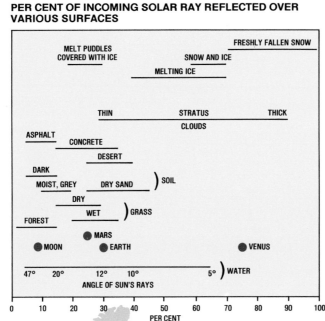

PER CENT OF INCOMING SOLAR RAY REFLECTED OVER VARIOUS SURFACES

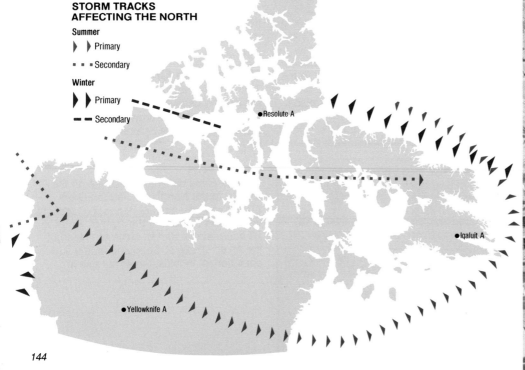

STORM TRACKS AFFECTING THE NORTH

Summer
- ▶ ▶ Primary
- ▪ ▪ ▪ Secondary

Winter
- ▶ ▶ Primary
- ▬ ▬ Secondary

● Resolute A

● Iqaluit A

● Yellowknife A

The winter circulation originates over northern Siberia, the Arctic ice-pack, and to some extent, the Greenland ice-cap. These dry Arctic systems can persist for weeks and effectively keep out incursions of contrasting warmer air and its associated bad weather from the west, south, and the east.

Pressure patterns change rapidly in spring and summer when the paths followed by the low pressure systems are generally diverted well north of their winter routes. Westerly currents from the Pacific, occurring at 3- or 4-day intervals, influence the mainland. There is an increased incidence of storms across the North, and these produce most of the summer's precipitation. Precipitation from local disturbances also occurs in the southern Mackenzie during July and August.

The only exception to the summer-storm/winter-calm pattern is found in the eastern Arctic. Here low pressure activity is frequent throughout the year and is most vigorous in fall and winter. July and August have few intense storms, but by October the storm activity has rejuvenated.

NORTHERN WATERS

The chilling Arctic waters, never clear of ice for very long, have a profound influence on all of Canada's climate. In the North, the maritime effect is felt most in summer and fall, and in many ways the feeling is unpleasant. Much of the region's fog, low cloud, and mist can be traced to the chilling open waters of the Arctic Ocean and to pools of water on permanent ice. Complete freeze-up is a slow process, which is usually

not completed until early December. In the meantime, open water in the eastern Arctic is an especially important precipitation source after mid- or late September and is partly the cause of very intense coastal snow squalls in October. Storms also make September and October generally the snowiest months of the year. The western Arctic, on the other hand, is much drier, and the northwest is strongly influenced by the adjacent multi-year ice cover of the Arctic Ocean.

Northern waters are covered with ice for much of the year. North of 75°N, complete clearing never really takes place, and it is only by August 1 that most of the ice cover dissipates in the Parry Channel and south to the mainland. The ice leaves northern Hudson Bay, Davis Strait, and the area east of the Beaufort Sea about three weeks earlier. In winter the ice-covered seas become almost an extension of the continent. Some moderation of winter air temperatures does occur throughout the ice-pack, but the difference amounts to only a few degrees, which, at -35 or -40°C, may be hard to notice.

Permanently frozen subsoil — permafrost — is found beneath the Beaufort Sea and across almost all the land area of the Northwest Territories. In high latitudes its thickness may exceed 500 m. Permafrost has a passive effect on climate. Water from melting snows or warm-season rains cannot percolate through the frozen layer, and the result is water ponding. Evaporation from these ponds uses up the energy that could otherwise warm the air. The active permafrost layer — that portion of the ground that

freezes in winter and thaws in summer — is thin, less than 1 m in depth. It becomes thinner, in response to the decrease in temperature, as one moves northward.

Not the Snowiest, Coldest, Windiest

The Northland is one of the driest regions in the world. Annual precipitation totals range from 200 to 500 mm across the mainland of the Territories. Precipitation reaches a definite peak in summer and early fall, but there is considerable variation from year to year. Across the western and central islands we find the real desert of Canada, with annual totals averaging less than 100 mm. However, because temperatures are so low, the evaporation rate is small and any precipitation seems excessive. On Baffin Island, over the high plateau facing Baffin Bay, yearly rain and snow totals exceed 300 mm, and on the highlands facing Davis Strait, they surpass 500 mm, largely as a result of frequent Atlantic coast storms, especially in the fall.

Across the south, 40 to 50% of the annual precipitation total occurs as rain over four months of the year, and in the very far north this percentage decreases to about 30%. Days with precipitation are more frequent than one would expect, given the light amounts. On the average, precipitation occurs one day in four, although along the southeast coast of Baffin Island, the frequency is one day in three. Rain may occur over the mainland from April to October and in the north for four months. Freezing rain can occur anywhere in any month.

ENVIRONMENT CANADA

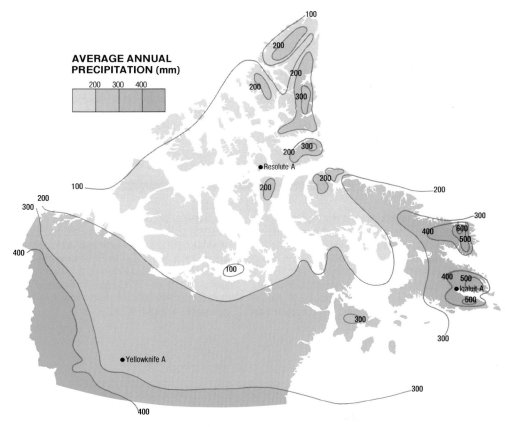

AVERAGE ANNUAL PRECIPITATION (mm)

200 300 400

It must be noted that it is difficult to measure precipitation accurately, because the rain and snow are deflected so much by the wind. In general, standard precipitation gauges err on the side of underestimating the amount of precipitation.

There is a noticeable southward increase in snowfall across the North. Total annual snowfalls range from 200 cm in the Mackenzie Valley, to 100 cm along the Arctic coast, to 60 to 100 cm on the northernmost islands. A typical snowfall total for much of

NORTHWEST TERRITORIES WEATHER SUPERLATIVES

	What	Where	What
Temperature			
highest	39.4°C	Fort Smith	July 18, 1941
lowest	-61.7°C	Fort Good Hope	Dec. 31, 1910
Greatest ...			
precipitation	1071 mm	Cape Dyer	1979
snowfall	997 cm	Cape Dyer	1978-79
number of fog days	253	Resolution Island	1963
wet period	201 days	Iqaluit	1981
number of blowing snow days	119	Baker Lake	1961-61
sunshine	2412 hours	Yellowknife	1976
wind (maximum hourly)	190 km/h	Resolution Island	Dec. 24, 1969
frost-free period	147 days	Yellowknife	1973

the Northwest Territories is 100 cm (compared with Montréal's 243 cm and Winnipeg's 125 cm).

Apart from the east coast of Baffin Island, heavy snowfall is a rarity in the North. Resolute's worst 24-hour snowfall was a meagre 13.2 cm on September 18, 1977, and Yellowknife has not had more than the 24 cm it received on February 20, 1982.

Across the Territories, snow falls on 50 to 100 days of the year and is likely to fall in any month, although there are places along the Mackenzie River where snow has never occurred in July and August.

Although snow amounts are relatively light, the snow is also very fine, and blows and drifts and lasts to a degree unknown in southern Canada. Snowcover usually first appears and finally disappears by September 1 and July 1, respectively, across the top of the islands, by September 15 and June 1 over the central islands and Baffin Island, and by late October and early May across the mainland in the Mackenzie Basin.

Snowcover dominates the northern climate. It covers the ground about 10 months of the year. Inland, it averages 30 cm in depth, but a vast portion of the barren lands in Arctic Canada accumulates relatively little snow. Depths increase toward the sea coasts and reach about 300 cm on exposed mountain slopes. Snowcover has

a vital role to play in protecting plant life and rodents from the cold and winds and in supplying them with water. On the other hand, it is an obstacle to grazing caribou.

Freezing Precipitation

Freezing precipitation in the North is about as common as it is over Western Canada, occurring about 5 to 10 days a year on average. Over half of the freezing precipitation falls during September and October with a secondary maximum in spring. Only along the southeastern coast of Baffin Island do frequencies exceed 15 days a year, brought on by incursions of warm, moist maritime air from the Atlantic Ocean.

Cold but Not Low

Canada's North is infamous as the source of the frigid Arctic air that rides over most of North America every winter and on occasion penetrates to the tip of the Florida Keys. For nearly nine months of the year, the average temperature across most of the Arctic does not rise above freezing. Even though the annual temperature over the Arctic Islands is lower than in any other part of Canada, it is still about 10°C higher than at sites along the same latitude in Soviet Siberia.

Eureka, on northern Ellesmere Island, has the distinction of being Canada's coldest weather station: the average annual temperature is -19.7°C. In February, the coldest month, Eureka has a mean monthly temperature of -38°C, and in February 1979 it recorded its coldest mean monthly temperature, a frigid -47°C.

In midwinter, ocean waters below the ice exert a slight moderating influence, so that often the lowest temperatures of the winter are found in the central areas of the large islands and over the mainland. Alert's all-time minimum is -50°C, compared with -58.3°C for Dawson, Yukon, and -57.2°C for Fort Smith, N.W.T.

Across the Northwest Territories, the highest mean winter temperature occurs in the Hudson Strait area, where mostly open water keeps midwinter averages in the -20 to -25°C range. The lowest winter averages, about -35°C, occur in the sheltered valleys and inlets north of 80°N and over the barrens of Keewatin, close to the central reservoir of

DEBUNKING NORTHERN MYTHS
Questions
(1) Does it ever rain in the Arctic?
(2) Can crops be grown north of the Arctic Circle?
(3) Are Arctic temperatures the lowest in Canada?
(4) Are snowfalls deep in the Arctic?
(5) Does the sun shine at midnight?
(6) Is the Arctic the windiest region in Canada?
(7) Do thunderstorms occur in the Arctic?
(8) Has a funnel cloud ever been reported north of the Arctic Circle?

Answers
(1) Yes. At our most northern weather stations, rain falls in four of the summer months.
(2) Where soils are suitable, coarse grains, vegetables, feed crops, and hay can be grown.
(3) On the average, yes, but the lowest temperatures ever recorded in Canada occurred outside the Northwest Territories.
(4) No. On the contrary, the snowfall is light. Victoria receives more snowfall than Eureka, N.W.T. An exception is Cape Dyer, where the average yearly snowfall amounts to 600 cm.
(5) Yes, for about 137 days between April 8 and September 6 north of 82°N latitude. In winter, for 135 days between October 16 and February 27 the noon hour is pitch black.
(6) Although it may seem so, especially with so much blowing snow, the Atlantic Provinces, Quebec, coastal British Columbia, and parts of Ontario and the Prairies actually have higher mean annual wind speeds.
(7) Yes.
(8) Yes.

cold Arctic air.

On occasion, mild air persists in winter as it did in January 1958 when the whole of the eastern Arctic experienced abnormal warmth, average temperatures being as much as 9°C above normal.

A striking feature of the Arctic climate is that temperature differences across the North are usually greater in summer than in winter. Western Arctic stations along the Mackenzie Valley have summer average temperatures that are warmer or only slightly cooler than those at many localities on the western Prairies. For example, the average July temperature at Fort Simpson is 16.6°C, while at Edmonton it is 15.8°C. Around the ice-filled waters of the Arctic Islands and along the northern fringe of the mainland, however, the summer is uniformly cooler. July temperatures here hover near

7.5°C and stay cool throughout the summer. The average July maximum at Resolute is 6.8°C, but at Fort Simpson it is 23.2°C. The cold waters of Hudson Bay also have a refrigerating effect on the adjacent shoreline. Summer temperatures along its coast, both day and night, average 3 to 5°C cooler than places inland.

Temperatures as high as 32°C have been reported from as far north as the Arctic coast, and as high as 36°C from the head-waters of the Mackenzie River. On the coastal areas of the northernmost islands, however, temperatures above 20°C are infrequent.

The combination of cold winters and cool summers gives the North enormous temperature ranges, commonly in the vicinity of 40°C, but it is in the Mackenzie Valley, where summers are warmer, that the nation's greatest temperature ranges are found. Fort Good Hope, for example, has an average range of 48°C.

Winter Discomfort

Severely high wind chill is a frequent and life-threatening hazard in the North, where it is not uncommon for air temperatures to fall below -40°C while winds blow at more than 30 km/h.

It is difficult to imagine a more penetrating, lasting cold than that reported at Iqaluit (Frobisher Bay) from February 6 to 18, 1979, when maximum winds of 60 km/h with gusts to 100 km/h accompanied extreme temperatures of -41°C for 12 days. Another extreme case occurred at Hall Beach on January 13, 1975, when the wind was clocked at 61 km/h in air temperatures that dipped to -46°C, producing a wind chill of 3188 W/m²(-86°C). On the mainland, Baker Lake and Chesterfield Inlet have the highest "1% January wind chill" values of all weather stations in

Canada, 2800 W/m²or an equivalent temperature of -71°C. In other words, this value is exceeded by only 1% of all January wind chill observations at these sites.

Frost

Across the north, frosts can occur during every month of the year, but in the south, where frost matters, it is rare from June through mid-August. The frost-free period varies from 50 to 100 days in the Mackenzie Basin — short, but adequate for most basic root or cold season crops — and from 40 to 60 days in the barrens west of Hudson Bay.

All or Nothing

Despite the potential for continuous or nearly continuous daylight in summer, the North is not a sunny region. Indeed, next to the Pacific northwest coast and Newfoundland, the central Arctic is the least sunny region in Canada. Annual sunshine totals range from a low of 1400 hours in Davis Strait and the central Arctic Islands to close to 2000 hours south of Great Slave Lake. For comparison, other Canadian average sunshine totals are Victoria 2059, Calgary 2314, Winnipeg 2321, Ottawa 2009, and Moncton 1929.

Given the long spell of winter darkness and the high frequency of low cloud and drizzle in the other seasons, a day with sunshine is often more appreciated by residents in the North than by their countryfolk in the South.

Thunderstorms

SOME OR NONE

Thunderstorms are not entirely unknown, but they are not frequent either, averaging less than one a year north of the Parry Channel. Mould Bay had only two occurrences in 30 years, and there are some places where thunder has never been heard. Across most of the Arctic, including Hudson Bay, an average of 1 to 5 thunderstorms occur each year. The Territories' maximum of 10 to 15 days occurs in the west, south of Great Slave Lake.

Restrictions

FOG AND BLINDING SNOW

With water everywhere, it is not surprising that the region has extensive cloud cover and frequent spells of fog.

Arctic sea fog occurs in summer when mild, moist southerly air travels northwards and is chilled while it crosses the ice-cold sea-water. Even the evaporation of water from saturated ground surfaces produces a cooling effect on the overriding air masses. Sea fog is dependent on the strength of the wind, being more frequent with light winds than with strong winds. Strong winds lift the

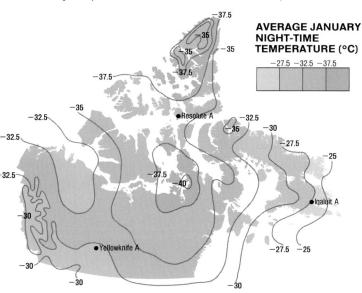

AVERAGE JANUARY NIGHT-TIME TEMPERATURE (°C)

−27.5 −32.5 −37.5

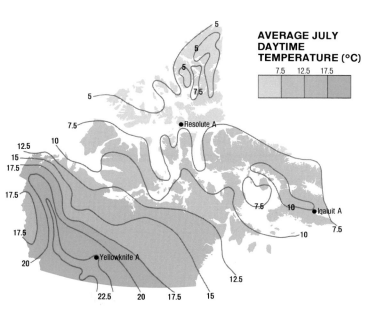

AVERAGE JULY DAYTIME TEMPERATURE (°C)

7.5 12.5 17.5

fog above the earth's surface, where it becomes visible as low, ragged stratus cloud. Sea fog is most prevalent between June and September and occurs everywhere along the coast, though it is most frequent in Hudson Strait, Davis Strait, and Baffin Bay.

From October through the open-water period of winter, steam fog or Arctic sea smoke forms when very cold air passes over areas of warmer open water. A slight warming of the air next to the water surface increases the rate of evaporation from the sea, but since the air above is too cold to absorb the extra moisture, it quickly condenses to form fog. Arctic sea smoke is highly localized and usually does not persist for more than a kilometre downwind. It also becomes less frequent as winter progresses and the ice cover grows, cutting off the moisture source.

Another winter fog form — ice fog — is a problem, especially when air temperatures dip below -40°C and wind speeds are under 10 km/h. Ice fog also forms at higher temperatures (-30°C) near settlements where industrial and transportation activity add moisture to the air. Increasing temperature and stronger winds break down the inversion layer which is required to maintain the ice fog.

Annually, fog days amount to 20 over the Mackenzie Valley, 50 around Hudson Bay, 60 to 90 just off the coast in the Arctic Ocean, and between 90 and 180 off the east coast of Baffin Island.

Wind

Living in the North is made both easier and more difficult by the prevailing wind conditions. Over water, wind moves ice, creates waves, and whips up freezing spray. On land, it rearranges snow, exposes food for grazing, and forces animals and humans to seek shelter. Wind-whipped blizzards make travelling even short distances hazardous. On the other hand, winds pack snow and clear fog, making movement easier.

Local topography strongly affects wind direction and speed. Coastal winds usually blow parallel to the shore. Prevailing winds in straits and fiords blow along the channel and increase in speed as the channel narrows.

Over land, winds tend to be about 30% lighter than winds at exposed coastal sites all year-round. Also wind speeds in the western Arctic are somewhat lower than winds in the east, where, in the Davis Strait, 2- or 3-day storms are common monthly occurrences from October to April.

Average values are somewhat misleading in the winter, since most stations report

ENVIRONMENT CANADA

SNOW BOX SCORE

	Average annual snowfall (cm)	Number of snowy days	Total Snowfall (cm) on snowiest day
Eureka	**44**	**46**	**15**
Resolute	**84**	**82**	**13**
Inuvik	**177**	**99**	**44**
Yellowknife	**135**	**82**	**28**
Baker Lake	**100**	**73**	**30**
Victoria	50	13	53
Regina	116	58	28
Montréal	235	62	41
Halifax	271	64	48
Windsor	117	45	37

AVERAGE ANNUAL NUMBER OF WET DAYS

● Snow
● Rain

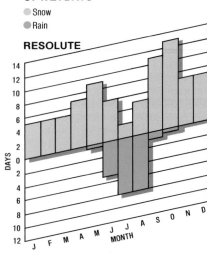

RESOLUTE

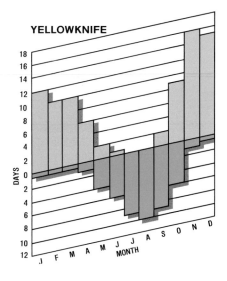

YELLOWKNIFE

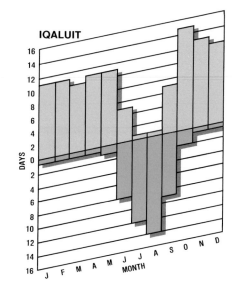

IQALUIT

calm conditions up to 30% of the time. Eureka reports 45% of its winter winds as calm, a typical figure for the northern Arctic Islands. Summer winds have far fewer calms but fewer gales also, and they tend to be more variable than in winter.

Blowing It Around

Although snow amounts in the Arctic are light, that which does fall is powder-fine and becomes easily airborne when the wind starts up. Wind speeds of 40 km/h or more are almost certain producers of blowing snow but do not always restrict visibilities to less than 1 km. On the other hand, winds of 60 km/h almost certainly lower visibility to under that limit. Blowing snow is the chief cause of restricted visibility in the North during the winter. Around Hudson Bay and to the north over the Arctic Islands, blowing snow accounts for at least one half of all the observations with restricted visibility. Baker Lake and Resolute each report an average of 91 days a year with blowing snow. Adding cold temperatures to the blowing snow conditions makes for the most notorious and vicious of winter storms — the blizzard.

During spring and fall, when the sun lies low in the sky, and the snow and sky are of a uniform whiteness, the landscape blends with the clouds so that the horizon disappears, forming a whiteout. During Arctic whiteouts, travel, even short distances, becomes extremely perilous, as it is easy to lose one's orientation and sense of direction.

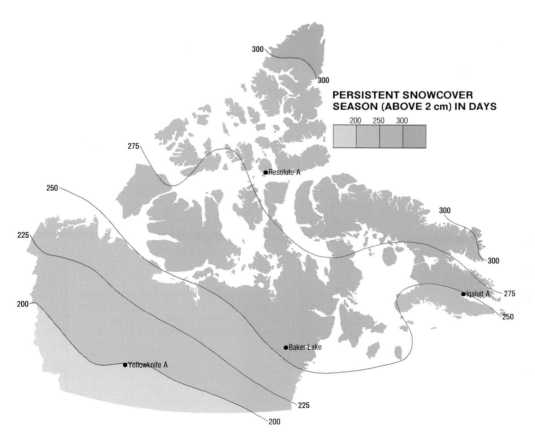

PERSISTENT SNOWCOVER SEASON (ABOVE 2 cm) IN DAYS

200 250 300

PERCENTAGE OCCURRENCES OF HOURS WITH BLOWING SNOW AND FOG BY MONTH

Blowing snow
Fog

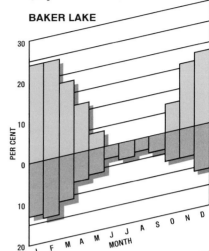

BAKER LAKE

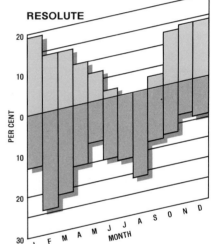

RESOLUTE

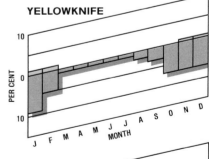

YELLOWKNIFE

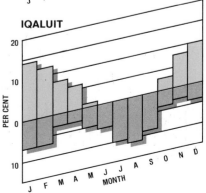

IQALUIT

Past and Future Climates

Not too long ago, many people considered climate to be constant and stable, interrupted every 100 000 years or so by an ice-age. The truth is that climate change is an intrinsic property of the "normal" climate and that throughout Canada's history, the climate has been unstable, shifting from extremely cold to mild conditions. Paleoclimatic information from sediment cores laid down in ancient lake bottoms, from rocks whose ages date back three billion years, from plant and animal fossils and from the land itself, clearly shows that Canada has fluctuated through warm and cold periods of relatively stable climate and prolonged, sometimes rapid, episodes of climate change that differed substantially from those of this century. At various times Canada has been covered by thick glaciers, lush tropical forests, freshwater lakes and salt-water seas. The evidence is irrefutable - salt mines in Windsor, coal beds in Nova Scotia, oil and gas deposits in Alberta, clay-bottom glacial lakes in Northern Ontario and cypress swamps on Ellesmere Island in the High Arctic.

ENVIRONMENT CANADA

TRADING CLIMATES
By the year 2030, enhanced greenhouse warming will give the following Canadian places the 1980s climate of the corresponding American cities.

Toronto	Indianapolis/Pittsburgh
Sudbury	Cleveland
Winnipeg	Minneapolis
Charlottetown	Boston
Edmonton	Cheyenne, Wyoming
Vancouver	San Francisco

GREENHOUSE EFFECT

The green house effect is no mystery. Nearly a hundred years ago, Swedish and American scientist independently advanced the hypothesis that changes in the abundance of carbon dioxide in the atmosphere would effect the surface temperature of the earth.

The atmosphere that blankets the earth acts like a gigantic greenhouse; the gases are like the glass letting the sunlight in but trapping the heat. The sun's visible rays pass readily to the earth's surface warming up the oceans and the land. Some of the sunlight is reflected directly to space by clouds, snow and ice. But the earth absorbs much of it, converting it into infrared energy or heat. The earth then sends these heat waves back out toward space. Some escape directly to space but most hit and are absorbed by molecules of carbon dioxide (CO_2) and other similar gases, setting them vibrating, spinning and waggling. Some heat energy is reflected back to earth, intensifying the warming effect.

What humankind is doing by increasing the atmospheric abundance of molecules of CO_2 and other greenhouse gases, such as freon gases used in refrigerators and aerosol sprays, methane from rice paddies and cattle, and nitrous oxide from fertilizers and the burning of fuel, is simply to increase the global temperature perhaps by about 2 to 5°C by the mid 20th century. Ironically, without the same greenhouse effect, the earth would be as barren, cold and lifeless as the surface of the moon. The heat-trapping blanket of naturally occurring gases maintains the average surface temperature of the Earth at a tolerable, near-room value of 16°C rather than a chilly -18°C. Mars has some CO_2 in its very thin, water-free atmosphere. Its surface temperature reaches -50°C at best. Venus is rich in CO_2, has an atmosphere 90 times denser than that of Earth, and has no oceans to absorb the gases; consequently its greenhouse effect maintains surface temperatures above 400°C.

Adding greenhouse gases to the atmosphere would upset the thermal balance between incoming and outgoing radiation. It is like throwing an extra blanket on an already comfortable body - body heat does not readily escape into the room, more heat is trapped beneath the blankets and the body becomes uncomfortably warm.

The earth's atmosphere contains substantially more CO_2 than it did before the Industrial Revolution. By analyzing cores from Greenland and Antarctic ice-caps, which enclose trapped bubbles of centuries-old gases, scientists have concluded that in 1750 the atmosphere contained about 280 parts per million of CO_2. Today the figure is about 350 ppm, nearly 25% higher. If that trend accelerates, at some point in the middle of the twenty-first century, perhaps as early as 2030, the combined effects of increased greenhouse gas concentrations will have a warming effect similar to concentrations of CO_2 almost double the pre-industrial value. Sooner rather than later, the greenhouse effect will produce climatic and ecological changes at a pace never before seen on Earth. It is these varying rates of climate change perhaps 100 times faster than at any time in human history that have ecologists worried. At deglaciation, the warming was about 1 to 2°C per 10 centuries; by comparison, projections for future warming range from a low of 2°C to 5°C in half a century. The bottom line to this means we are changing the system much faster than any natural change.

WHAT IS THE GREENHOUSE EFFECT?
Sunlight passes through the atmosphere and warms the Earth.

Earth re-radiates infrared heat energy — some escapes to space, but water vapour, dust, carbon dioxide (CO_2) freons, nitrous oxide, methane and other greenhouse gases in the atmosphere absorb and trap most of the heat, returning it to the Earth.

Higher concentrations of greenhouse gases trap more heat; the atmosphere and earth's surface temperatures increase.

Burned forests release excessive CO_2 and other gases to the atmosphere, preventing the earth's infrared (heat) energy from escaping into space thus heating up the Earth even more.

CO_2 and other industrial gases from automobiles and factories remain trapped in the atmosphere where they surround the Earth like a blanket.

Climate Change

In describing climate change, it is important to distinguish long-term from short-term change. Many studies of climate change focus on change over millenia or eons of time and attempt to explain the causes of the ice-age. Proxy sources - pollen, tree-rings, ice cores, lake sediments - are often used to reconstruct past climates for such studies.

Instrumental records are often used to document short-term changes, especially over the past 100 years. Short-term variations are generally of a much shorter duration and much greater amplitude and intensity than long-term trends. They are also of greater economic-social importance because they lead to famines, energy shortages, inflation and migration.

Weather Records

The instrumental record in Canada dates from 1840 when the British Government established a Magnetic and Meteorological Observatory at Toronto. The Toronto record has continued for 150 years within one kilometre of the original site. During the 1870s, while the weather service in Canada was being organized, the weather station network expanded across Canada. Data exist in 1990 for over a century in parts of every province and for nearly 80 years in parts of northern Canada. In 1871 there were 125 observing stations in Canada and in 1990 there were over 2,500.

A study of the trends in temperature averaged over successive five-year periods at several localities shows decidedly colder years at the beginning of this century, warming in the 1920s through to the 1950s, cooling in the sixties, stabilizing in the 1970s and since 1979 trending to warmer temperatures. The cool period in the 1960s was most pronounced in eastern Canada. Since precipitation varies greatly within a region, an analysis of national trends is less revealing.

Changes in climate arise in part from such diverse factors as fluctuations in the earth's orbit around the sun (from a near circle to a marked ellipse and back again); variations in the tilt of the earth's axis of rotation (changes of 3 ° can cause a shift in the poles such as happened 50 000 years ago when the North Pole was located in what is now Hudson Bay); changes in solar (sunspot) energy output (although strong evidence exists for an approximate 11-year cycle of sunspots, the direct link between this cycle and changes in rainfall, drought, temperatures and stormy weather is unsubstantiated); drifting of the continents; mountain building and volcanic activity (which injects dust, ash and gases that

CO₂ TRENDS — LAST 300 YEARS

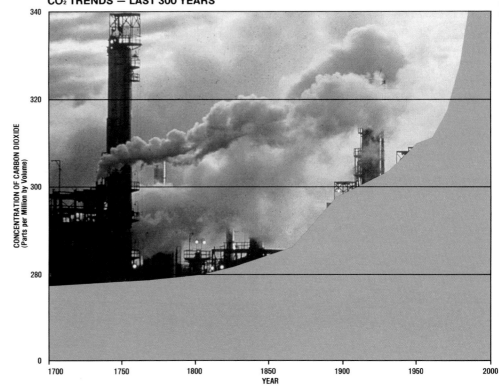

increase the reflection of the incoming solar radiation before it can hit the earth, thus cooling the atmosphere); the presence of carbon dioxide and other greenhouse gases; and man's activities (such as massive land-use changes - deforestation and urbanization - and large water diversion projects).

Future Climate

Studies of future climate, based on complex computer simulations, indicate that the global concentrations of the CO_2 and more than 20 other greenhouse gases are expected to double those of pre-industrial periods sometime before the middle of the next century.

Possible changes to the Canadian climate that could result by the year 2050 include the following:

- In general, winters would be 3 to 5°C warmer in the south and 10 to 15°C - warmer in the Arctic; summers would be warmer by 2°C in the south and 3°C in the Arctic.

- Storm tracks would be displaced northward increasing Arctic precipitation; snow seasons would be shorter in the Arctic but snow totals and spring runoff might be greater.

- Slow melting of the permafrost would lead to instability.

- Ice-free seasons would be longer everywhere, with possibly an ice-free Arctic summer but increased iceberg hazards.

- Drier weather in southern areas could lead to more frequent and severe droughts.

- Wetter conditions in British Columbia may trigger more mudslides, avalanches and flooding.

- In Atlantic and Pacific Canada, sea-levels are likely to rise, which would lead to coastal flooding.

- The duration of the growing season and the amount of available summer heat would increase substantially.

- A global warming of 2 to 5°C might generate a larger frequency of extreme events like droughts and floods and lead to more intense thunderstorms, hurricanes and tornadoes.

- Great Lakes water levels would drop (by close to 1 m) under warmer climates.

- In the south, heat waves would be longer and more frequent.

Impacts of a Changed Climate

Needless to say, the effects of climate change on Canada's economy, society and environment are expected to be substantial, both in threatening and beneficial ways.

Changes would be most evident in the resource sectors — agriculture, water and energy. Concerns exist about water shortages, reduced soil moisture, falling lake levels, rising sea-levels, reduced river flow and worsening droughts. At the same time, climate change might offer easier navigation on the Great Lakes and in the Arctic, lower heating bills, lengthen the growing season, and induce milder winters and longer summers.

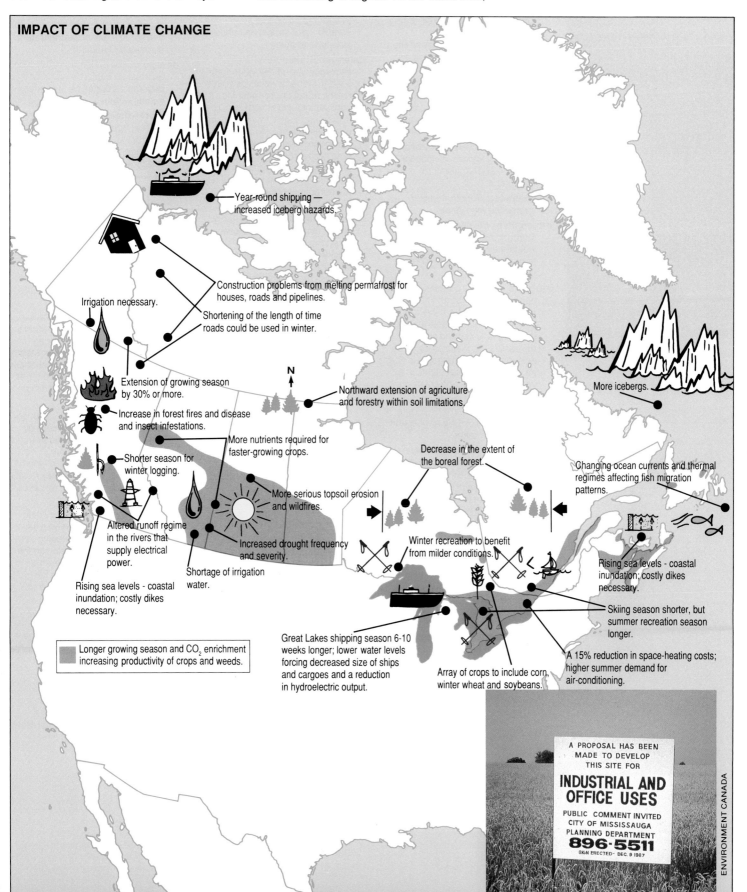

IMPACT OF CLIMATE CHANGE

Year-round shipping — increased iceberg hazards.

Construction problems from melting permafrost for houses, roads and pipelines.

Irrigation necessary.

Shortening of the length of time roads could be used in winter.

Extension of growing season by 30% or more.

More icebergs.

Increase in forest fires and disease and insect infestations.

Northward extension of agriculture and forestry within soil limitations.

More nutrients required for faster-growing crops.

Decrease in the extent of the boreal forest.

Changing ocean currents and thermal regimes affecting fish migration patterns.

Shorter season for winter logging.

More serious topsoil erosion and wildfires.

Altered runoff regime in the rivers that supply electrical power.

Increased drought frequency and severity.

Winter recreation to benefit from milder conditions.

Rising sea levels - coastal inundation; costly dikes necessary.

Shortage of irrigation water.

Rising sea levels - coastal inundation; costly dikes necessary.

Skiing season shorter, but summer recreation season longer.

Longer growing season and CO_2 enrichment increasing productivity of crops and weeds.

Great Lakes shipping season 6-10 weeks longer; lower water levels forcing decreased size of ships and cargoes and a reduction in hydroelectric output.

Array of crops to include corn, winter wheat and soybeans.

A 15% reduction in space-heating costs; higher summer demand for air-conditioning.

A PROPOSAL HAS BEEN MADE TO DEVELOP THIS SITE FOR

INDUSTRIAL AND OFFICE USES

PUBLIC COMMENT INVITED
CITY OF MISSISSAUGA
PLANNING DEPARTMENT
896-5511

SIGN ERECTED - DEC. 9 1987

ENVIRONMENT CANADA

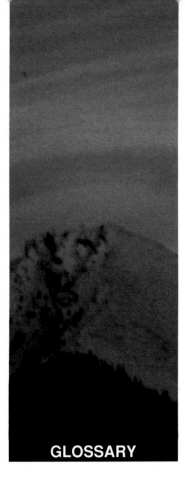

GLOSSARY

acid precipitation: the deposition of strong acids from the atmosphere in the form of rain, snow, fog or dry particles. The acid in the rain is the result of pollution caused primarily by the discharge of sulphur oxides and nitrogen oxides into the atmosphere from the burning of coal and oil, during the operation of smelting industries and from transportation. In the atmosphere these gases combine with water vapour to form acids.

advection: the transport of a property or constituent of the air such as temperature or moisture solely by the motion of the atmosphere.

air mass: an extensive body of air with a fairly uniform distribution of moisture and temperature throughout.

anemometer: an instrument for measuring wind speed.

atmosphere: the envelope of air surrounding the earth. Most weather events are confined to the troposphere, the lower 10 km of the atmosphere.

aurora (e.g., aurora borealis, northern lights): the luminous, radiant emission from the upper atmosphere over middle and high latitudes, and centred on each of the earth's magnetic poles. This shimmering, pulsating light is often seen on clear winter nights in a variety of shapes and colours.

average annual range: the difference between the average temperatures of the warmest and the coldest months.

blizzard: severe winter weather condition characterized by low temperatures, strong winds above 40 km/h, and poor visibilities less than 1 km due to blowing snow and lasting three hours or more.

blocking high: a nearly stationary anticyclone that inhibits the west to east migration of cyclones.

blowing snow day: a day when the horizontal visibility is restricted to 10 km or less because snow particles are raised by the wind to a height of 2 m or more. Blowing snow is higher than drifting snow (under 2 m).

break up: the breaking of ice that normally begins when air temperatures rise significantly above 0°C, and the ice begins to melt. The process is aided by the action of winds and currents and results in mechanical breaking of the ice. Break up ends when the water body becomes completely free of all ice.

bright sunshine: sunshine intense enough to burn a mark on recording paper mounted in the Campbell-Stokes sunshine recorder. The daily period of bright sunshine is less than that of visible sunshine because the sun's rays are not intense enough to burn the paper just after sunrise, near sunset and under cloudy conditions.

Campbell-Stokes sunshine recorder: the official instrument used in Canada since the 1880s for recording bright sunshine. It consists of a glass sphere 10 cm in diameter that is mounted in part of a spherical bowl to which a cardboard card is affixed. The card is scorched by the sun's rays and simple measurements of the length of the scorch marks gives the number of hours of bright sunshine for the day.

chinook (also snow-eater): a dry, warm, strong wind that blows down the eastern slopes of the Rocky Mountains in North America. The warmth and dryness are due principally to heating by compression as the air descends the mountain slope.

climate: the synthesis of day-to-day weather variations in a locality. The climate of a specified area is represented by the statistical collection of its weather conditions during a specified interval of time, which is usually taken to include the following weather elements: temperature, precipitation, humidity, sunshine, and wind velocity.

climate controls: fairly permanent factors that shape the nature of climate in a locality. They include solar radiation, atmospheric circulation, topography, distribution of land and water, and land use, e.g., urban, rural and forest.

climate severity index: a number scale from 0 to 100 that describes many of the unfavourable (uncomfortable, depressing, confining and hazardous) aspects of the climate - their intensity, duration and frequency of occurrence.

cold wave: an occurrence of dangerously cold conditions that usually lasts longer than a few days when temperatures often dip below -18°C.

condensation: the conversion of water vapour into water droplets in the form of fog, clouds, or dew; the opposite of evaporation.

daily range of temperature: the difference between the daily lowest temperature (usually occurring just before midnight or around sunrise) and the highest temperature (usually around midafternoon).

degree-day: the difference between the average temperature for a day and some reference temperature, such as 5°C and 18°C.

dew-point: the temperature to which air must be cooled in order for it to become saturated by the water vapour within.

diffuse radiation: radiant energy coming from different directions in contrast to (direct) parallel radiation. Examples include radiation scattered and reflected by water vapour, clouds and other atmosphere gases.

drought: an extended period of dry weather that lasts longer than expected or than normal and leads to measurable losses (crop damage, water supply shortage).

dust devil: a small, vigorous, well-developed whirlwind made visible by the dust, sand or other debris picked up from the ground.

equinox: the moment when the sun passes directly above the earth's equator. The vernal or spring equinox occurs on or about March 21 and the fall equinox on or about September 22. On these days all over the earth the days and nights are of equal duration.

evaporation/evapotranspiration (of water): the physical process by which water is transferred from the earth's surface to the atmosphere through the evaporation of water or ice into water vapour, and through transpiration from plants.

evaporation pan: an open, cylindrical metal pan mounted on a flat wooden base. Daily readings of the water level in the pan and the amount of precipitation from a nearby rain gauge enable the amount of evaporation to be calculated.

fast ice: ice attached to the shore, beached, stranded in shallow water, or frozen to the bottom of shallow water bodies (sea, river, lake).

flash floods: a very rapid rise of water with little or no advance warning, most often when an intense thunderstorm drops a huge rainfall on a fairly small area in a very short space of time.

fog day: a day when fog occurs, i.e., when the horizontal visibility is restricted to less than 1 km (thick fog) or 10 km (fog) because very small water droplets are suspended in the air.

freeze-up: ice formation in rivers, lakes and coastal water bodies. The process begins when the surface water cools below 0°C and ice crystals begin to form. It ends when the water body has attained its maximum ice coverage.

freezing precipitation: supercooled water drops of drizzle, or rain, which freeze on impact to form a coating of ice upon the ground and on the objects they strike.

freezing precipitation day: a day with an occurrence of freezing rain or freezing drizzle of any quantity, even a trace, that freezes on impact.

freezing spray: icing caused by the freezing of sea spray. A combination of low temperatures and strong winds causes sea spray to freeze on structures at sea (usually limited to 15-20 m above the sea surface) or near the water's edge. It is the most widely reported and usually the most serious form of icing at sea.

front: the interface or boundary between two different air masses which have originated from widely separated regions. A cold front is the leading edge of an advancing cold air mass, while a warm front is the trailing edge of a retreating cold air mass.

frost: the deposit of ice crystals that occurs when the air temperature is at or below the freezing point of water. The term frost is also used to describe the icy deposits of water vapour that may form on the ground or on other surfaces like car windshields, which are colder than the surrounding air and which have a temperature below freezing.

frost-free season: the period between the last spring frost and the first fall frost.

frost hollow: a depression in the surface topography conducive to the accumulation of cold air at night as a result of the downslope flow of colder, denser air beneath warmer, lighter air.

gale: a strong wind. A gale warning is issued for expected winds of 65 to 100 km/h (34 to 47 knots).

global radiation: the total solar radiation coming from the sun and all of the sky, including the direct beam from the unclouded sun as well as the diffuse radiation arriving indirectly as scattered or reflected sunlight from the sky and clouds.

greatest absolute range of temperature: the difference between the all-time extreme high and low temperatures.

greenhouse effect: the warming of the atmosphere as a result of the net absorption and re-emission in the atmosphere of infrared radiation principally by carbon dioxide gas.

growing season: the part of the year when the growth of natural and cultivated vegetation is made possible by sufficiently high temperatures, usually when the average daily temperature remains above 5°C.

gust: a sudden, brief increase in wind speed, for generally less than 20 seconds.

hail day: day with an occurrence of precipitation of small lumps or pieces of ice (hailstones) with a diameter of 5 mm or more. Hail sizes usually range from that of small peas to that of cherries, but have been observed as large as oranges.

hailstorm-alley: a region of high hailstorm frequency in the Calgary-Drumheller-Red Deer triangle of central Alberta. An average of 68 hail days per year occurs annually in the "alley".

heat island: the region of warm air over a city where temperatures are higher than in surrounding rural areas.

heat wave: a period with more than three consecutive days of maximum temperatures at or above 32°C.

high (also called an anticyclone): a term for an area of high (maximum) pressure with a closed, clockwise (in the Northern Hemisphere) circulation of air.

humidex: a measure of what hot weather "feels like". Air of a given temperature and moisture content is equated in comfort to air with a higher temperature and that of negligible moisture content. At a humidex of 30°C some people begin to experience discomfort.

hummock: broken ice that has been forced upwards by pressure.

hurricane (also typhoon, willy-willy): a tropical storm with wind speeds of 120 km/h (65 knots) or more that can be many thousands of square kilometres in size. They originate over the warm tropical ocean as a small low-pressure system. They have a life span of several days and occur more frequently from August to October.

ice cover: the ratio of an area of ice of any concentration to the total area of water surface within some geographic area, e.g., a bay with 6/10 surface ice cover has about 6 parts ice to 4 parts open water.

ice fog (also frost fog and frozen fog): a fog composed of suspended ice crystals. It forms at temperatures below -30°C when water vapour condenses directly into tiny ice crystals in clear and calm weather.

ice island: a large piece of floating ice about 5 m above sea-level that has broken away from an Arctic ice shelf, having a thickness of 30-50 m and an area ranging from a few thousand square metres to 250 square kilometres or more, and usually characterized by a regular undulating surface that gives it a ribbed appearance from the air.

ice pellets: precipitation consisting of fragments of ice, 5 mm or less in diameter, that bounce when hitting a hard surface making a sound upon impact.

ice ridge: a line or wall of broken ice forced up by pressure. It may be fresh or weathered.

Indian summer: a warm calm spell of weather with clear skies, sunny but hazy days, and cool nights, occurring in fall, reminiscent of the balmier days of summer. Some people believe that at least one frost and abnormally cool weather must precede a true Indian summer.

inversion: a temperature increase with height that is a departure from the usual decrease with height; may also refer to moisture.

isobar: a line on a weather map or chart connecting points of equal atmospheric pressure.

January thaw (also bonspiel thaw): a period of mild weather with above freezing temperatures supposed to recur each January and lasting anywhere from a few hours to a week or month.

killing frost: a frost severe enough to end the growing season, usually when the air temperature falls below -2°C.

lake-effect snow (also sea-effect snow): snow generated by the passage of cold air over relatively warm lake, sea or river water. In the Great Lakes region, copious snow may fall over terrain in the snowbelts for some distance downwind of the lakes.

lake evaporation: evaporation from small natural water bodies, such as small lakes, reservoirs and ponds, in which the heat stored by their water mass is negligible.

lapse rate: the rate of decrease of an atmospheric variable with height, e.g., temperature or moisture.

low (also cyclone): an area of low (minimum) atmospheric pressure that has a closed counter-clockwise circulation in the Northern Hemisphere.

moisture deficit or deficiency: the difference between the potential evapotranspiration, e.g., water losses under conditions where moisture supplies are not lacking, and the actual water loss evaporated from the vegetation and the ground.

Nipher snow gauge: a collector used at principal stations to measure the water equivalent of snowfall by melting it.

orographic effect: the lifting of air over a barrier such as a mountain. The lifted air cools, moisture condenses and precipitation falls usually on the upslope surface. It may extend for some distance downwind of the barrier. Generally the lee side is the dry rain-shadow.

pack ice (ice-pack): an area of sea ice. Compact, consolidated, close, and open are used to describe the concentration of ice, i.e. the ratio describing the mean areal density of ice.

peak wind (gust): the highest instantaneous wind speed recorded for a specific time period.

permafrost: a layer of permanently frozen soil and rock. The active layer refers to that portion of the ground that freezes in winter and thaws in summer, usually less than 1 m in depth.

polar easterlies: shallow easterly winds located in the higher latitudes north of 55°N.

polar night: a period longer than 24 hours during which the sun stays totally below the horizon, e.g., at Resolute polar night lasts from November 9 to February 2.

possible sunshine: the ratio (expressed as a per cent) of the duration of sunshine that is observed to the duration of sunshine that would be observed if the terrain were level, no obstructions existed, and the sky were clear.

precipitation: any and all forms of water, whether liquid or solid, that fall from the atmosphere and reach the earth's surface. A day with measurable precipitation is a day when the water equivalent of the precipitation is equal to or greater than 0.2 mm.

prevailing wind direction: the wind direction most frequently observed during a specific time period.

rain-shadow: (see orographic effect).

radiation cooling: the cooling of the earth's surface and nearby air that results when the surface loses more heat than it gains.

reflectivity (albedo): the ratio of the amount of radiation reflected by a body to the amount of radiation incident upon it, commonly expressed as a percentage.

relative humidity: the ratio of the actual amount of water vapour in the air to the amount present if the air were saturated at the same temperature.

sea smoke (also Arctic sea smoke, sea mist, water smoke, steam fog): fog formed when very cold air is situated above relatively warm water. Commonly observed over lakes and rivers on cold fall mornings.

significant wave height: the mean height of the highest third of all waves observed in a wave train.

silver thaw: euphemism for freezing rain or glaze storm, frequently used in Newfoundland where such storms are often at their maximum intensity.

smoke or haze day: a day when the visibility is restricted to 10 km or less because small dry particles are suspended in the air. These particles must be sufficiently numerous to give the air a greenish-yellow, light-greyish or bluish hue for smoke, or a milky or pearly hue for haze.

snow: precipitation consisting of white or translucent ice crystals and often agglomerated into snowflakes. A day with measurable snow is a day when the total snowfall is at least 0.2 cm.

snowbelt: area where the prevailing onshore winds are responsible for heavy snowfall, usually downwind of open water.

snowcover: the accumulation on the ground of snowfall, ice pellets, and frost glaze. The snowcover season begins when 2.5 cm or more of snow remains for 7 days, and ends when less than 2.5 cm of snow is on the ground in spring.

snow roller: jelly-roll or muff-looking snow cylinders that occur in snow-covered fields; produced following several centimetres of sticky and wet snow and during strong gusty winds and freezing temperatures. Ideal surface conditions for the formation of snow rollers include a sloping terrain with an older, crusted snowcover.

solstice: the time at which the noon-day sun is at its farthest distance north or south of the celestial equator. The days are longest and shortest in the Northern Hemisphere at the summer solstice (about June 21), and the winter solstice (about December 21), respectively.

storm surge (also storm tide): a large rise of water along a shore as a result, mostly, of storm winds.

storm track: the path taken by a low-pressure centre.

stratus cloud: low-level cloud in the form of a grey layer with a rather uniform base; precipitation is not usually associated with stratus.

thunderstorm: a local storm, usually produced by a cumulonimbus cloud, and always accompanied by thunder and lightning. A thunderstorm day is a day when thunder is heard or when lightning is seen (rain and snow need not have fallen).

tornado (also twister): a violently rotating column of air that is usually visible as a funnel cloud hanging from dark thunderstorm or cumulonimbus clouds. It is one of the least extensive of all storms, but in violence, it is the most destructive.

tropic of Cancer: the northern parallel of latitude that is the northernmost latitude (23°27'N) where the sun is directly overhead at noon.

tropic of Capricorn: the southern parallel of latitude that is the southernmost latitude (23°27'S) where the sun is directly overhead at noon.

tsunami: waves generated by underwater earthquakes.

twilight: the interval of incomplete darkness following sunset and preceding sunrise whose duration varies with latitude and date.

typhoon: a severe tropical cyclone in the Pacific Ocean; counterpart of the Atlantic hurricane.

vapour pressure: the pressure exerted by water vapour in the atmosphere.

weather: state or condition of the atmosphere with respect to heat or cold, wetness or dryness, calm or storm, and clearness or cloudiness for a certain period of time.

westerlies (west-wind belt): the pronounced west-to-east motion of the atmosphere centred over middle latitudes from about 35 to 65° latitude.

white-out: a uniformly white glow in Arctic or northern regions over an unbroken snowcover and beneath a uniformly overcast sky. Neither shadows, horizon nor clouds are discernible; the sense of depth and orientation is lost.

wind chill: a simple measure of the chilling effect experienced by the human body when strong winds are combined with freezing temperatures. The larger the wind chill, the faster the rate of cooling. The wind chill factor is expressed in watts per square metre or in °C (an equivalent temperature).

wind direction: the direction from which the wind is blowing.

FURTHER READINGS

General Sources

C.C. Boughner and M.K. Thomas, *The Climate of Canada.* In: Canada Year Book, Ottawa. 1959.

Energy, Mines and Resources Canada, *The National Atlas of Canada.* 5th Edition. 1985.

Environment Canada, *Canadian Climate Normals 1951-1980.* 9 volumes. 1982-1984.

Environment Canada, *Climatic Atlas Climatique-Canada.* 7 map series. 1984 to 1989.

Environment Canada, *Hydrological Atlas of Canada.* 34 maps. 1978.

F.K. Hare and J. Hay, *The Climate of Canada and Alaska.* In: World Survey of Climatology. Elsevier, Amsterdam. 1974.

F.K. Hare and M.K. Thomas, *Climate Canada.* 2nd Edition. J. Wiley and Sons. 1979.

Climate Controls

R.A. Bryson and F.K. Hare, *The Climates of North America.* Volume 11. World Survey of Climatology. Elsevier, Amsterdam. 1974.

J.B. Maxwell, *The Climate of the Canadian Arctic Islands and Adjacent Waters.* Envir. Canada (AES). Volume 1. 1980.

D.F. Putnam, *Canadian Regions - A Geography of Canada.* Dent. 1952.

Reader's Digest Association (Canada) Ltd., Atlas of Canada, 1981.

Temperature

Environment Canada, *Canadian Climate Normals 1951-1980.* 1979.

Environment Canada, *Climatic Atlas Climatique-Canada.* Map Series 1. 1984.

Environment Canada, *Hydrological Atlas of Canada.* Plates 14 and 15. 1978.

J.M. Masterton and F.A. Richardson, *Humidex.* Envir. Canada (AES). 1979.

M.K. Thomas and D.W. Boyd, *Wind Chill in Northern Canada.* The Canadian Geographer. 1957.

Rain And Snow

S.A. Changnon, *The Climatology of Hail in North America.* Meteorological Monographs. Volume 16. December 1977.

Environment Canada, *Canada Water Year Book.* Six issues. 1975 to 1985.

Environment Canada, *Canadian Climate Normals 1951-1980.* Volume 3. 1982.

Environment Canada, *Climatic Atlas Climatique - Canada.* Map Series 2. 1986.

Environment Canada, *Downwind - The Acid Rain Story.* 1981.

Environment Canada, *Hydrological Atlas of Canada.* Plates 3, 9, 10 and 11. 1978.

Sunshine Country

Environment Canada, *Canadian Climate Normals 1951-1980.* Volume 1 and Volume 7. 1982.

Environment Canada, *Climatic Atlas Climatique - Canada.* Map Series 4. 1987.

Environment Canada, *Hydrological Atlas of Canada.* Plate 12. 1978.

J.E. Hay, *An Analysis of Solar Radiation Data for Selected Locations in Canada.* Envir. Canada (AES). 1977.

D.C. McKay and R.J. Morris, *Solar Radiation Data Analyses for Canada 1967-1976.* 6 Volumes. Envir. Canada (AES). 1985.

Something In The Wind

Tim Christison, *Snow-Eater.* In: Nature Canada. Spring 1986.

Environment Canada, *Canadian Climate Normals 1951-1980.* Volume 5, 1982.

Environment Canada, *Climatic Atlas Climatique - Canada.* Map Series 5, 1988.

Environment Canada, *Hydrological Atlas of Canada.* Plate 16. 1978.

F. Forrester, *Winds of the World.* Weatherwise. October 1982.

Fog... The Thick of It

M.O. Berry and R.G. Lawford, *Low-temperature Fog in the Northwest Territories.* Envir. Canada(AES). 1977.

Environment Canada, *Canadian Climate Normals 1951-1980.* Volume 9, 1984.

M.J. Newark, *Fog.* In: The Canadian Encyclopedia. Hurtig Publishers. 1985.

Ice... The Thin of It

W.T.R. Allen, *Freeze-up, Break-up and Ice Thickness in Canada.* Envir. Canada (AES). 1977.

Environment Canada, *Hydrological Atlas of Canada.* Plate 19. 1978.

W.E. Markham, *Ice Atlas : Canadian Arctic Waterways.* Envir. Canada (AES). 1981.

W.E. Markham, *Ice Atlas Eastern Canadian Seaboard.* Envir. Canada (AES). 1980.

W.E. Markham, *Ice Atlas - Hudson Bay and Approaches, Envir. Canada (AES). 1987.*

It's The Humidity

Environment Canada, Canadian Climate Normals 1951-1980, Volume 8, 1984.

Environment Canada, Climatic Atlas Climatique - Canada. Map Series 3, 1987.

J.M. Masterton and F.A. Richardson, *Humidex.* Envir. Canada (AES). 1979.

D.W. Phillips and R.B. Crowe, *Climate Severity Index for Canadians.* Envir. Canada (AES). 1982.

S.E. Tuller, *Mean Monthly and Annual Precipitable Water Vapour in Canada.* Weather. July 1972.

Evaporation

Environment Canada, Canadian Climate Normals 1951-1980. Volume 9, 1984.

Environment Canada, *Hydrological Atlas of Canada.* Plates 17. 1978.

H.L. Ferguson et. al., *Mean Evaporation over Canada.* Water Resources Research. December 1970.

Floods - Too Much Water

Environment Canada, *Canada Water Year Book.* Six issues. 1975 to 1985.

D.W. Phillips and G.A. McKay (editors), *Canadian Climate In Review 1980.* Envir. Canada (AES). 1981.

Drought - Too Little Water

Alberta Environment, *Drought.* In: Environment News. Spring 1986.

AES Drought Study Group, *An Applied Climatology of Drought in the Prairie Provinces.* Envir. Canada (AES). 1986.

D.W. Phillips and G.A. McKa y (editors),*Canadian Climate In Review 1980.* Envir. Canada (AES). 1981.

Stormy Weather

Environment Canada, *Climatic Atlas Climatique* -Canada. Map Series 3. 1987.

J.L. Knox, The Storm "*Hazel*". Bulletin American Meteorol. Society. June 1955.

M.J. Newark, *Tornadoes in Canada 1950 to 1979*. Envir. Canada (AES). 1983.

Newfoundland and Labrador

J.A. Peach, *The Tourism and Outdoor Recreation Climate of Newfoundland and Labrador*. Envir. Canada (AES). 1975.

Dept. of Transport (Meteorol. Branch), *The Climate of Quebec and Labrador*. 1953.

Prince Edward Island

A.D. Gates. *The Tourism and Outdoor Recreation Climate of the Maritime Provinces*, Envir. Canada (AES). 1975.

L. Jacobs. *The Meteorology of the Gulf of St. Lawrence and Prince Edward Island*. British Air Ministry Office. 1945.

W. Simpson et al. *Gulf of St. Lawrence Water Uses and Related Activities*. Envir. Canada (LANDS). 1973.

Nova Scotia

A.D. Gates. *The Tourism and Outdoor Recreation Climate of the Maritime Provinces*, Envir. Canada (AES). 1975.

N.S. Dept. of Development, *Nova Scotia Development Atlas - Climate*. 1973.

W.G. Richards and J.F. Amirault, *The Climate of Sable Island* (draft manuscript). Envir. Canada (AES). 1986.

New Brunswick

E.E.D. Day et. al., *The Climate of Fundy National Park and Its Implications for Recreation and Park Management*. Saint Mary's University, Dept. of Geography. 1977.

A.D. Gates. *The Tourism and Outdoor Recreation Climate of the Maritime Provinces*, Envir. Canada (AES). 1975.

Quebec

Dept. of Transport (Meteorol. Branch), *The Climate of Quebec and Labrador*. 1953.

Environment Canada, *The Climate of Montreal*. Atmospheric Environment Service. 1987.

M.G. Ferland and R.M. Gagnon, *Le Climat du Québec Meridional*. Québec : Ministère des Richesses Naturelles. 1967.

N.N. Powe, *The Climate of Montreal*. Dept. of Transport (Meteorol. Branch). 1969.

C.V. Wilson, *The Climate of Québec* (2 volumes). Envir. Canada (AES). 1971 and 1973.

R.M. Gagnon and M.G. Ferland, *Le Climat du Québec Septentrional*. Quebec: ministère des Richesses naturelles.

A. Houde's Atlas Climatologique du Québec: Température, Précipitation. Ministère de l'Environnement, Direction de la Météorologie 1978.

Ontario

T. Allsopp et. al., *Climatic Variability and Its Impact on the Province of Ontario*. Envir. Canada (AES). 1981.

D.M. Brown et. al., *The Climate of Southern Ontario*. Envir. Canada (AES). 1980.

L.J. Chapman and M.K. Thomas, *The Climate of Northern Ontario*. Envir. Canada (AES). 1968.

R.B. Crowe, *The Climate of Ottawa-Hull*. Envir. Canada (AES). 1983.

R.B. Crowe, et. al., *The Tourist and Outdoor Recreation Climate of Ontario*. (4 volumes). Envir. Canada (AES). 1973.

P. Globus and S. Robertson, *Ontario's Benign, Consistent, Reliable Weather?* Leisure Ways. 1985.

J.M. Masterton and D.W. McNichol, *A Recreational Climatology of the National Capital Region*. Envir. Canada (AES). 1981.

D.W. Phillips and J.A.W. McCulloch, *The Climate of the Great Lakes Basin*. Envir. Canada (AES). 1972.

A. Saulesleja, Great Lakes Climatological Atlas Envir. Canada. (AES). 1986.

Manitoba

W.G. Kendrew and B.W. Currie, *The Climate of Central Canada*. Queen's Printer. 1955.

R. Longley, *The Climate of the Prairie Provinces*. Envir. Canada (AES). 1972.

J.M. Masterton, et. al., *The Tourism and Outdoor Recreation Climate of the Prairie Provinces*. Envir. Canada (AES). 1976.

R. Raddatz et. al., *Manitoba and Saskatchewan Tornado Days 1960 to 1982*. Envir. Canada (AES). 1983.

Saskatchewan

W.G. Kendrew and B.W. Currie, *The Climate of Central Canada*. Queen's Printer. 1955.

R. Longley, *The Climate of the Prairie Provinces*. Envir. Canada (AES). 1972.

J.M. Masterton, et. al., *The Tourism and Outdoor Recreation Climate of the Prairie Provinces*. Envir. Canada (AES). 1976.

G.A. McKay, et. al., *The Agricultural Climate of Saskatchewan*. Dept. of Transport (Meteorol. Branch). 1967.

R. Raddatz et. al., *Manitoba and Saskatchewan Tornado Days 1960 to 1982*. Envir. Canada (AES). 1983.

Alberta

T. Christison, *Snow-eater*. In: Nature Canada. Spring 1986.

B. Janz and D. Storr, *The Climate of the Contiguous Mountain Parks*. Envir. Canada (AES). 1977.

P. Klivokiotis and R.B. Thomson, *The Climate of Calgary*. Envir. Canada (AES). 1986.

R. Longley, *The Climate of the Prairie Provinces*. Envir. Canada (AES). 1972.

R. Longley, *Climate and Weather Patterns*. In Alberta : A Natural History. W.G. Hardy (editor) Chapter 3. 1967.

J.M. Masterton et. al., *The Tourism and Outdoor Recreation Climate of the Prairie Provinces*. Envir. Canada (AES). 1976.

Rod Olson, *The Climate of Edmonton*. Envir. Canada (AES). 1985.

British Columbia

R. Chilton, *A Summary of Climatic Regimes of British Columbia*. BC Ministry of Environment. 1981.

J. Hay and T. Oke, *The Climate of Vancouver*. Tantalus Research Ltd (Vancouver). 1976.

W.G. Kendrew and D. Kerr, *The Climate of British Columbia and the Yukon Territory*. Dept. of Transport (Meteorol. Branch). 1956.

S.E. Tuller, *Climate*. In: Vancouver Island: Land of Contrasts, Univ. of Victoria, Dept. of Geography. 1979.

Yukon

F.J. Eley and B.F. Findlay, *Agroclimatic Capability of Southern Portions of the Yukon Territory and Mackenzie District*, NWT. Envir. Canada (AES). 1977.

W.G. Kendrew and D. Kerr, *The Climate of British Columbia and the Yukon Territory*. Dept of Transport (Meteorol. Branch). 1955.

H. Wahl, et. al., *Climate of Yukon*. Envir. Canada (AES). 1987.

Northwest Territories

B.M. Burns, *The Climate of the Mackenzie Valley-Beaufort Sea*. (2 volumes). Envir. Canada (AES). 1973 and 1974.

Canadian Hydrographic Service, *The Climate of the Canadian Arctic*. Dept. of Energy, Mines and Resources. 1970.

J.B. Maxwell, *The Climate of the Canadian Arctic Islands and Adjacent Waters*. (2 volumes). Envir. Canada (AES). 1980 and 1982.

J.B. Maxwell, *A Climate Overview of the Canadian Inland Seas*. In: Canadian Inland Seas (I.P. Martini, editor), Elsevier, Amsterdam, 1986.

Past and Future Climates

J. P. Bruce and H. Hengeveld, *Our Changing Northern Climate*. GEOS (Energy, Mines and Resources). 1985.

Environment Canada, *Understanding CO_2 and Climate*. Annual Reports 1984-1988.

H.L. Ferguson and D.W. Phillips, *Developing A National Climate Program*. Canadian Climate Program (AES). 1986.

R. Revelle, *Carbon Dioxide and World Climate*. Scientific American. 1982.

M.K. Thomas, *Recent Fluctuations in Canada*. Envir. Canada (AES). 1975.

United Nations Environment Programme, *The Greenhouse Gases*. 1987.

CLIMATE DATA

Climate Tables

The following pages contain climate data from a selection of 17 weather stations across Canada. The tables contain monthly and annual data for temperature, precipitation, humidity, degree-days, sunshine, and "number of days with" for eight meteorological elements, and sunrise and sunset times. Other information includes climatic extremes and frost data and a brief description of the local area in terms of terrain, water bodies, and other topographical features. The latitude, longitude, and altitude of each location appear.

Tables are shown for airport sites as representative places within each of the different climatic regions of Canada.

The following notes are provided to clarify the information included in the tables.

Notes

Two basic statistics are presented. *Average*, mean, or normal refer to the value of the particular element averaged over the period from 1951-1980. *Extremes* are based on the full period of years in which observations have been made under conditions with essentially the same exposure. Extremes generally cover all available data to the end of 1985.

- a "**day with**" is counted only once per day regardless of the number of individual occurrences of the phenomenon during that day. "**Days with**" include the following:
  - **gale winds** exceed 63 km/h
  - **hail** is an ice piece with a diameter of 5 mm or more
  - **thunder** is reported when thunder is heard or lightning or hail is seen
  - **fog** is a suspension of very small water droplets in the air that reduces the horizontal visibility at eye level to less than one kilometre
  - **freezing temperature** is a temperature below 0.0°C
  - **freezing precipitation** is rain or drizzle of any quantity, even a trace, that freezes on impact
  - **rain** is a measurable amount of liquid water (rain, showers, or drizzle) equal to or greater than 0.2 mm.
  - **snow** is a measurable amount of solid precipitation (snow, snow grains, ice crystals, or ice and snow pellets) equal to or greater than 0.2 cm.
- **bright sunshine** is in hours and tenths
- each degree of mean temperature below 18°C is counted as one **heating degree-day**; above 18°C, one **cooling degree-day**; and above 5°C, one **growing degree-day**. Monthly and annual sums are calculated for each period and averaged over 1951-1980 to establish the normal.
- **prevailing direction** (with its percentage of occurrence) is the most common single wind direction (direction from which the wind blows) without regard for wind speed. **Average speed** is the average of all winds and calms.
- **Extreme gust speed** is the peak wind on record, observed from a dial indicator or abstracted from a continuous chart record.
- **average snow on ground** is for the last day of the month
- **total precipitation** is the total water equivalent of snowfall and rainfall
- **frost** is said to occur if the temperature falls to 0.0°C or lower. Average dates are given for the last spring frost and first fall frost. The period between these two dates is the average length of the frost-free season.

Abbreviations

*	denotes a value less than 0.5 (but not zero)
M	missing
E	estimate
PM	polar night
MS	midnight sun

St. John's, Newfoundland

LONG-TERM AVERAGES AND EXTREMES OF CLIMATE DATA

	TEMPERATURE (°C)					AVERAGE NUMBER OF DAYS WITH								BRIGHT SUNSHINE	SUNRISE/SUNSET
	AVERAGE			EXTREMES						Freezing					
	Midafternoon	Early Morning	Daily	Highest	Lowest	Winds (>63 km/h)	Hail	Thunder	Fog	Temp.	Precip.	Rain	Snow	(hours)	Local Standard Time (h:min) on the 15th of Month
Jan	-0.5	-7.2	-3.9	15.2	-23.3	5	0	*	8	30	8	10	18	70.6	7:45 16.35
Feb	-1.0	-7.9	-4.5	12.8	-23.3	3	0	*	8	27	7	8	16	83.4	7:08 17.22
Mar	0.9	-5.5	-2.3	18.3	-20.6	4	0	*	11	29	9	10	15	94.6	6:16 18.04
Apr	4.5	-2.2	1.2	21.7	-14.8	1	0	*	13	25	8	12	10	115.5	5:14 18.49
May	9.7	1.1	5.4	25.6	-6.7	1	0	*	16	14	1	15	3	158.9	4:24 19:30
Jun	15.8	5.9	10.9	29.4	-3.3	*	*	1	13	1	0	13	*	186.6	4:02 20.00
Jul	20.2	10.7	15.5	30.6	-1.1	*	0	1	14	*	0	13	0	220.1	4:17 19.56
Aug	19.4	11.1	15.3	30.0	0.6	*	0	1	11	0	0	15	0	186.0	4:55 19.16
Sep	15.6	7.5	11.6	27.8	-1.1	1	*	*	7	*	0	15	*	146.5	5:37 18.16
Oct	10.4	3.4	6.9	22.8	-5.6	1	*	*	8	6	0	18	2	110.4	6:18 17.16
Nov	6.5	0.2	3.4	19.4	-12.6	3	*	*	8	16	1	16	7	68.0	7:05 16.26
Dec	1.6	-4.6	-1.5	16.1	-19.4	4	0	*	7	28	4	11	17	56.8	7:42 16.09
Ann	8.6	1.0	4.8	30.6	-23.3	23	*	3	124	176	38	156	88	1497.4	

	HUMIDITY			DEGREE-DAYS (°C)			WIND				PRECIPITATION					
	Relative (%)		Vapour Pressure kPa								RAINFALL (mm)		SNOWFALL (cm)			
	6 a.m. Average	3 p.m. Average		Heating (<18°)	Cooling (>18°)	Growing (>5°)	Prevailing Direction	% Frequency	Average Speed (km/h)	Extreme Gust (km/h)	Average	Wettest Day	Average	Snowiest Day	Average Snow on Ground	Total (mm)
Jan	86	82	0.42	678.0	0.0	1.0	W	18.7	27.5	167	77.9	84.6	81.4	38.4	21	155.8
Feb	85	80	0.40	635.4	0.0	0.5	W	17.6	27.5	193	69.7	67.1	74.6	54.9	25	140.1
Mar	87	79	0.46	630.1	0.0	1.6	W	14.2	26.9	148	67.0	53.3	65.0	45.7	14	131.9
Apr	88	76	0.56	504.6	0.0	6.0	WSW	14.5	24.4	159	78.1	91.7	34.6	31.6	T	115.6
May	88	73	0.73	390.6	0.0	53.9	WSW	17.3	22.9	146	89.4	50.8	11.1	25.4	0	101.8
Jun	87	70	1.04	216.0	1.8	180.1	WSW	23.4	22.2	108	83.4	75.2	2.0	13.5	0	85.6
Jul	88	69	1.39	93.1	14.6	324.5	WSW	26.2	21.4	107	75.3	121.2	0.0	T	0	75.3
Aug	90	72	1.41	95.3	11.7	319.5	WSW	23.8	21.2	113	121.6	80.5	0.0	0.0	0	121.6
Sep	88	72	1.13	193.5	0.9	197.8	WSW	19.0	22.1	122	116.7	66.0	T	0.3	0	116.7
Oct	88	76	0.87	343.6	0.1	78.0	WSW	16.6	23.8	122	140.9	100.8	4.4	19.8	T	145.5
Nov	87	81	0.69	439.1	0.0	28.5	W	14.1	25.2	145	140.6	69.9	21.2	25.3	2	162.5
Dec	85	82	0.50	605.1	0.0	4.7	W	17.2	26.8	153	96.7	85.1	65.1	49.3	13	161.2
Ann			0.80	4824.4	29.1	1196.1	WSW	16.6	24.3	193	1157.3	121.2	359.4	54.9		1513.6

AVERAGE FROST DATA

Last frost (spring) May 1
First frost (fall) October 11
Frost-free period 131 days

CLIMATIC EXTREMES

Wettest Month on Record	356.9 mm	December 1909
Highest Temperature on Record	30.6°	July 7, 1949
Lowest Temperature on Record	-27.2°	February 24, 1961

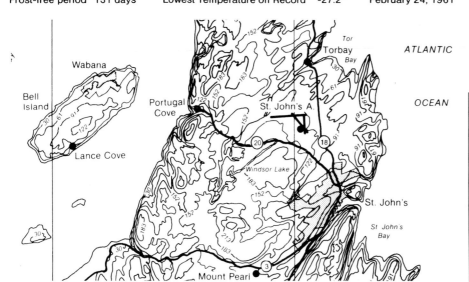

STATION INFORMATION

ST. JOHN'S   47°37'N   52°45'W   140 m

The airport is situated on the Avalon Peninsula, 7 km north-northwest of the city of St. John's and 5 km south of Torbay. The ocean and a rugged terrain with numerous small lakes and rivers dominate the surrounding landscape. The Atlantic Ocean lies 7 km east of the site while Conception Bay is about 10 km to the west. In all but the eastern quadrant, there are a large number of hills within a 3 to 8 km distance of the airport. The land peaks at a height of over 210 m near the tip of the peninsula about 18 km north of the site.

Charlottetown, P.E.I.

LONG-TERM AVERAGES AND EXTREMES OF CLIMATE DATA

	TEMPERATURE (°C)					AVERAGE NUMBER OF DAYS WITH								BRIGHT SUNSHINE	SUNRISE/SUNSET
	AVERAGE			EXTREMES						Freezing					
	Midafternoon	Early Morning	Daily	Highest	Lowest	Winds (> 63 km/h)	Hail	Thunder	Fog	Temp.	Precip.	Rain	Snow	(hours)	Local Standard Time (h:min) on the 15th of Month
Jan	-3.0	-11.2	-7.1	11.7	-27.8	1	*	*	4	30	4	7	15	90.0	7:52 — 16:51
Feb	-3.2	-11.6	-7.5	13.3	-27.8	1	*	*	3	28	3	4	13	113.6	7:17 — 17:36
Mar	0.6	-6.8	-3.1	16.1	-23.9	1	0	*	6	30	5	6	12	134.9	6:27 — 18:16
Apr	6.2	-1.6	2.3	22.2	-12.2	1	*	*	5	23	2	10	6	156.6	5:28 — 18.58
May	13.4	3.6	8.5	31.2	-6.7	*	*	1	5	6	*	14	1	196.6	4:41 — 19:37
Jun	19.4	9.5	14.5	32.2	-1.1	*	*	2	4	*	0	12	0	220.8	4:20 — 20.05
Jul	23.0	13.7	18.3	33.9	4.4	*	*	3	4	0	0	12	0	241.3	4:35 — 20.02
Aug	22.3	13.3	17.8	34.4	4.4	0	0	2	3	0	*	12	0	218.1	5:10 — 19.24
Sep	17.9	9.1	13.5	29.4	-0.6	*	*	1	2	*	0	12	0	176.0	5:49 — 18.27
Oct	12.0	4.2	8.1	23.3	-6.7	*	*	*	4	5	*	14	1	128.3	6:28 — 17.29
Nov	6.2	-0.5	2.9	18.9	-14.9	1	*	0	3	18	3	13	5	77.6	7:12 — 16.42
Dec	-0.2	-7.6	-3.9	16.7	-28.1	1	*	*	4	29	3	8	15	64.6	7:48 — 16.27
Ann	9.6	1.2	5.4	34.4	-28.1	6	*	9	47	169	17	124	68	1818.4	

	HUMIDITY			DEGREE-DAYS (°C)			WIND				PRECIPITATION					
	Relative (%)		Vapour Pressure kPa								RAINFALL (mm)		SNOWFALL (cm)			
	6 a.m. Average	3 p.m. Average		Heating (<18°)	Cooling (>18°)	Growing (>5°)	Prevailing Direction	% Frequency	Average Speed (km/h)	Extreme Gust (km/h)	Average	Wettest Day	Average	Snowiest Day	Average Snow on Ground	Total (mm)
Jan	83	79	0.35	778.6	0.0	0.2	W	21.2	22.1	142	42.7	74.4	76.8	47.2	26	116.8
Feb	83	76	0.32	719.5	0.0	0.1	W	18.5	20.8	132	32.8	53.1	65.8	47.5	32	97.4
Mar	85	76	0.43	653.9	0.0	1.1	W	11.6	21.7	129	31.8	35.1	61.6	33.8	18	95.3
Apr	86	69	0.56	471.2	0.0	14.7	N	11.1	19.9	121	53.9	58.7	27.3	38.1	1	81.8
May	86	64	0.84	295.5	0.5	118.8	WSW	10.2	18.9	116	81.3	53.3	2.1	13.2	0	83.6
Jun	87	65	1.27	117.1	10.7	283.6	WSW	14.9	18.0	97	79.9	53.3	0.0	T	0	79.9
Jul	89	66	1.64	28.8	39.1	413.3	WSW	16.4	16.1	130	84.3	70.0	0.0	0.0	0	84.3
Aug	90	66	1.60	37.8	32.6	397.8	WSW	16.6	16.0	87	88.1	76.7	0.0	0.0	0	88.1
Sep	89	67	1.26	139.5	5.7	256.2	WSW	13.1	17.2	121	86.3	56.1	0.0	T	0	86.3
Oct	87	70	0.91	306.5	0.0	110.1	W	15.9	19.1	137	103.8	106.4	2.6	21.6	0	106.4
Nov	86	77	0.67	456.8	0.0	26.8	W	19.4	20.4	116	97.4	51.1	21.6	30.5	0	120.5
Dec	85	81	0.43	683.4	0.0	3.2	W	21.4	21.5	177	58.8	50.4	72.8	32.0	2	129.0
Ann			0.86	4688.6	88.6	1625.9	W	13.5	19.3	177	841.1	106.4	330.6	47.5	20	1169.4

AVERAGE FROST DATA
Last frost (spring) May 16
First frost (fall) October 15
Frost-free period 151 days

CLIMATIC EXTREMES
Wettest Month on Record 315.0 mm September 1942
Highest Temperature on Record 34.4° August 12, 1944
Lowest Temperature on Record -30.5° January 18, 1982

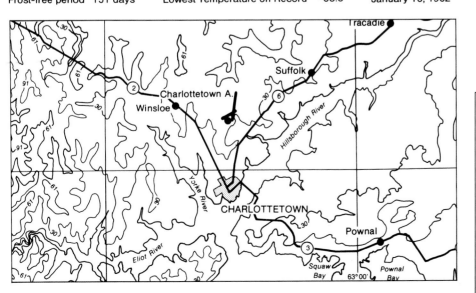

STATION INFORMATION

CHARLOTTETOWN   46°17'N   63°08'W   55 m
The airport is 5.6 km north of the centre of the City of Charlottetown, on the south shore of Prince Edward Island. The terrain surrounding the airport is rolling. The ground slopes to the south, southwest, north and southeast with level ground in all other directions. There are few wooded areas near the site and most of the land is devoted to agriculture.

Halifax, Nova Scotia
LONG-TERM AVERAGES AND EXTREMES OF CLIMATE DATA

| | TEMPERATURE (°C) | | | | | AVERAGE NUMBER OF DAYS WITH | | | | | | | | BRIGHT SUNSHINE | SUNRISE/SUNSET |
| | AVERAGE | | | EXTREMES | | | | | | Freezing | | | | | |
	Midafternoon	Early Morning	Daily	Highest	Lowest	Winds (>63 km/h)	Hail	Thunder	Fog	Temp.	Precip.	Rain	Snow	(hours)	Local Standard Time (h:min) on the 15th of Month
Jan	-1.6	-10.3	-6.0	13.5	-25.6	1	*	*	6	30	4	8	15	92.8	7:48 16:58
Feb	-1.6	-10.6	-6.1	12.2	-26.1	1	0	*	5	27	4	6	13	118.1	7:16 17:41
Mar	2.6	-5.9	-1.6	15.9	-18.3	*	*	*	8	29	5	8	11	139.8	6:28 18:18
Apr	7.8	-1.1	3.3	24.9	-9.6	*	*	*	10	21	2	10	6	164.9	5:32 18:57
May	14.5	3.8	9.2	32.8	-4.4	*	0	1	12	4	*	14	1	206.6	4:47 19:34
Jun	20.3	9.1	14.8	32.8	0.6	*	*	2	15	0	0	12	0	203.4	4:28 20:00
Jul	23.3	13.0	18.2	33.9	6.1	*	*	2	18	0	0	11	0	226.4	4:42 19:58
Aug	23.1	13.1	18.1	34.5	4.4	0	*	2	15	0	0	10	0	216.6	5:15 19:22
Sep	18.7	8.9	13.8	32.2	-0.6	*	*	1	9	*	0	10	1	182.2	5:52 18:27
Oct	13.0	4.1	8.6	25.6	-6.7	*	*	1	9	5	*	12	4	154.4	6:28 17:32
Nov	7.0	-0.3	3.4	19.4	-13.1	*	0	*	8	19	*	13	13	95.3	7:09 16:48
Dec	1.3	-7.0	-2.9	15.0	-23.3	1	0	*	7	28	4	11		84.5	7:43 16:34
Ann	10.7	1.4	6.1	34.5	-26.1	3	*	9	122	163	19	125	64	1885.0	

| | HUMIDITY | | | DEGREE-DAYS (°C) | | | WIND | | | | PRECIPITATION | | | | | |
| | Relative (%) | | Vapour Pressure kPa | | | | | | | | RAINFALL (mm) | | SNOWFALL (cm) | | | |
	6 a.m. Average	3 p.m. Average		Heating (<18°)	Cooling (>18°)	Growing (>5°)	Prevailing Direction	% Frequency	Average Speed (km/h)	Extreme Gust (km/h)	Average	Wettest Day	Average	Snowiest Day	Average Snow on Ground	Total (mm)
Jan	84	77	0.38	743.1	0.0	0.9	WNW	12.7	20.2	117	91.0	94.1	63.1	43.7	13	152.8
Feb	82	74	0.35	680.1	0.0	0.5	WNW	12.3	19.7	127	70.0	66.8	65.6	47.2	17	133.5
Mar	85	71	0.46	609.0	0.0	3.0	N	11.5	20.7	126	83.2	89.2	45.5	24.6	4	128.4
Apr	87	64	0.58	440.7	0.0	19.8	N	11.3	19.2	111	89.3	76.7	24.1	28.4	1	114.8
May	82	62	0.85	274.7	0.9	134.6	S	13.4	18.5	92	102.8	75.9	3.4	26.9	0	106.4
Jun	88	64	1.25	106.6	9.8	293.3	S	18.1	17.3	97	89.4	64.0	0.0	T	0	89.4
Jul	92	65	1.61	29.3	35.1	408.8	SSW	19.7	15.9	130	94.2	57.9	0.0	0.0	0	94.2
Aug	92	64	1.60	32.8	36.0	406.1	S	17.6	15.4	85	111.3	218.2	0.0	0.0	0	111.3
Sep	92	65	1.28	131.3	6.1	264.8	SSW	13.6	15.7	93	93.7	59.4	0.0	T	0	93.7
Oct	90	67	0.95	292.7	0.1	122.4	SSW	11.8	17.2	109	129.9	66.8	3.6	38.6	0	133.5
Nov	89	75	0.70	437.8	0.0	34.5	N	9.5	18.5	111	140.7	62.7	11.9	20.3	1	152.5
Dec	86	79	0.46	646.6	0.0	5.1	NW	11.8	19.8	132	128.3	98.8	53.9	47.5	10	180.1
Ann			0.87	4424.7	88.0	1693.8	SSW	11.2	18.2	132	1223.8	218.2	271.0	47.5		1490.6

AVERAGE FROST DATA

Last frost (spring) May 12
First frost (fall) October 15
Frost-free period 155 days

CLIMATIC EXTREMES

Wettest Month on Record	387.1 mm	August 1971
Highest Temperature on Record	34.4°	August 22, 1976
Lowest Temperature on Record	-26.1°	February 4, 1971

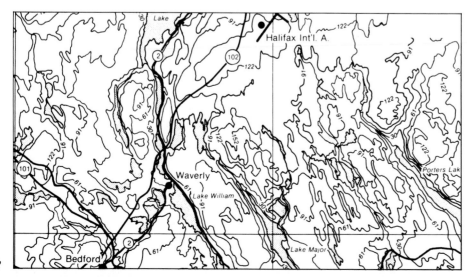

STATION INFORMATION
HALIFAX INT'L   44°53'N   63°31'W   145 m
Halifax Int'l Airport is situated on a ridge running in an east-northeast to west-southwest direction. The land is generally forested and slopes toward the Atlantic Ocean, 30 km to the south and Cobequid Bay 48 km to the north. The centre of the city of Halifax is about 27 km south-southwest and Shearwater A is about 27 km south of the site. The area beyond the immediate vicinity of the airport is characterized by rolling terrain. Small lakes are found in all directions from the site but are more predominant to the south and the southeast. Observations taken at the site are considered to be representative of the surrounding rural area.

Saint John, New Brunswick

LONG-TERM AVERAGES AND EXTREMES OF CLIMATE DATA

	TEMPERATURE (°C)					AVERAGE NUMBER OF DAYS WITH								BRIGHT SUNSHINE	SUNRISE/SUNSET
	AVERAGE			EXTREMES						Freezing						
	Midafternoon	Early Morning	Daily	Highest	Lowest	Winds (>63 km/h)	Hail	Thunder	Fog	Temp.	Precip.	Rain	Snow	(hours)	Local Standard Time (h:min) on the 15th of Month	
Jan	-2.6	-12.9	-7.8	14.0	-31.7	1	*	*	3	30	3	6	13	105.9	8:00	17:05
Feb	-2.2	-12.8	-7.5	11.1	-36.7	1	0	*	2	28	3	5	12	124.8	7:27	17:49
Mar	2.1	-7.1	-2.5	13.9	-30.0	1	*	*	5	29	3	7	11	143.5	6:38	18:28
Apr	7.9	-1.5	3.2	22.8	-16.7	1	*	*	7	22	1	10	5	157.6	5:40	19.08
May	14.4	3.6	9.0	30.0	-7.8	*	*	1	9	5	0	13	1	203.4	4:54	19:45
Jun	19.2	8.4	13.8	31.7	-2.2	0	*	3	14	*	0	14	0	203.0	4:35	20.13
Jul	22.2	11.6	16.9	32.8	1.1	*	*	3	19	0	0	12	0	218.5	4:49	20.10
Aug	21.7	11.4	16.6	34.5	-0.6	*	*	2	17	0	0	12	0	212.5	5:23	19.33
Sep	17.6	7.7	12.7	28.9	-6.7	*	*	1	13	1	*	12	0	166.0	6:01	18.37
Oct	12.1	3.1	7.6	25.6	-10.6	*	*	1	9	9	0	12	1	140.8	6:38	17.41
Nov	6.3	-1.6	2.3	21.7	-15.3	1	*	*	5	21	*	13	4	97.3	7:21	16.55
Dec	-0.1	-9.3	-4.8	16.1	-30.9	1	0	*	3	28	2	8	12	92.0	7:55	16.41
Ann	9.9	0.1	5.0	34.5	-36.7	6	*	11	106	173	12	124	59	1865.3		

	HUMIDITY			DEGREE-DAYS (°C)			WIND				PRECIPITATION					
	Relative (%)		Vapour Pressure kPa								RAINFALL (mm)		SNOWFALL (cm)			Total (mm)
	6 a.m. Average	3 p.m. Average		Heating (<18°)	Cooling (>18°)	Growing (>5°)	Prevailing Direction	% Frequency	Average Speed (km/h)	Extreme Gust (km/h)	Average	Wettest Day	Average	Snowiest Day	Average Snow on Ground	
Jan	80	72	0.33	798.7	0.0	0.5	NW	13.9	20.6	143	74.8	83.0	76.3	42.4	20	148.8
Feb	78	68	0.32	721.3	0.0	0.2	NW	13.9	20.2	146	53.7	82.3	63.2	34.8	22	115.7
Mar	82	67	0.42	635.3	0.0	1.3	NW	11.1	21.0	137	62.3	74.0	49.9	40.1	7	114.1
Apr	85	61	0.57	443.7	0.0	17.8	N	9.9	19.1	121	86.0	125.5	20.7	26.2	T	107.3
May	87	60	0.85	278.1	0.0	128.7	SSW	12.2	18.3	132	105.7	66.5	2.0	10.2	0	107.7
Jun	90	65	1.21	128.4	3.4	265.0	SSW	16.5	17.3	129	94.2	54.1	0.0	0.0	0	94.2
Jul	93	67	1.50	48.4	13.9	368.5	S	20.1	15.5	105	103.4	72.6	0.0	T	0	103.4
Aug	93	66	1.48	55.8	12.7	359.8	S	16.7	15.0	93	102.0	125.2	0.0	0.0	0	102.0
Sep	93	66	1.21	162.0	1.8	229.9	SSW	13.4	16.4	130	111.8	74.9	0.0	0.0	0	111.8
Oct	89	67	0.89	321.2	0.0	99.6	SSW	9.1	18.3	138	125.1	85.3	2.5	19.8	0	127.7
Nov	87	72	0.64	470.0	0.0	24.6	NW	11.0	19.8	126	131.2	154.4	14.5	21.3	2	145.7
Dec	83	75	0.40	705.2	0.0	3.5	NW	14.2	20.5	145	101.8	85.1	63.6	58.2	14	166.0
Ann			0.82	4768.1	31.9	1499.4	S	9.7	18.5	146	1152.0	154.4	292.7	58.2		1444.4

AVERAGE FROST DATA
Last frost (spring)   May 16
First frost (fall)   October 3
Frost-free period   139 days

CLIMATIC EXTREMES
Wettest Month on Record	317.5 mm	July 1950
Highest Temperature on Record	34.4°	August 22, 1976
Lowest Temperature on Record	-36.7°	February 11, 1948

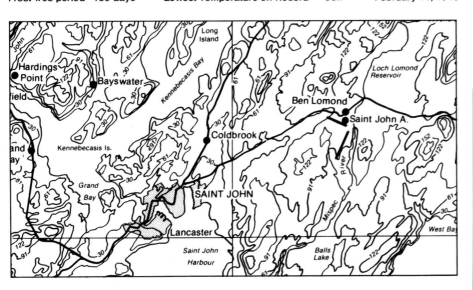

STATION INFORMATION
SAINT JOHN   45°19'N   65°53'W   109 m
The airport is located about 17 km east-northeast of downtown Saint John and the mouth of the Saint John River. The site is virtually surrounded by bodies of water. The surrounding countryside is essentially rolling and wooded with considerable ridging to the northeast, east and southeast, approximately 17 km from the site. The highest peak in this area is that of Upham Mountain, elevation 320 m, 23 km northeast of the airport.

Sept-Îles, Québec

LONG-TERM AVERAGES AND EXTREMES OF CLIMATE DATA

	TEMPERATURE (°C)					AVERAGE NUMBER OF DAYS WITH								BRIGHT SUNSHINE	SUNRISE/SUNSET	
	AVERAGE			EXTREMES						Freezing						
	Midafternoon	Early Morning	Daily	Highest	Lowest	Winds (>63 km/h)	Hail	Thunder	Fog	Temp.	Precip.	Rain	Snow	(hours)	Local Standard Time (h:min) on the 15th of Month	
Jan	-8.6	-19.3	-14.0	10.0	-43.3	2	*	0	1	31	1	2	15	107.8	7:19	15:49
Feb	-6.9	-18.0	-12.5	10.6	-38.3	1	0	0	2	28	1	1	12	137.9	6:38	16:41
Mar	-1.5	-11.7	-6.6	11.7	-31.7	1	*	*	3	30	2	2	11	153.3	5:41	17:28
Apr	4.1	-4.1	0.0	14.4	-24.4	1	*	*	3	26	1	6	7	186.7	4:34	18:17
May	10.4	1.3	5.9	28.3	-11.7	1	0	1	5	11	*	11	1	230.8	3:39	19:03
Jun	16.7	6.8	11.7	32.2	-2.8	*	0	2	6	*	0	13	*	234.3	3:14	19:36
Jul	19.8	10.7	15.2	32.2	1.7	*	*	2	8	0	*	13	0	242.9	3:30	19:31
Aug	18.8	9.3	14.1	31.1	-0.6	*	*	2	7	0	*	13	0	222.5	4:13	18:47
Sep	13.8	4.6	9.3	29.4	-6.5	*	*	*	5	4	*	13	0	157.8	4:59	17:42
Oct	7.8	-0.6	3.6	22.2	-12.8	1	*	0	5	18	*	11	0	125.8	5:46	16:36
Nov	1.4	-6.3	-2.5	16.9	-28.9	1	0	*	4	27	1	6	9	93.6	6:37	15:42
Dec	-6.2	-15.7	-11.0	9.4	-36.1	1	0	0	2	31	2	2	14	97.2	7:17	15:22
Ann	5.8	-3.6	1.1	32.2	-43.3	9	*	7	51	206	8	93	72	1990.6		

	HUMIDITY			DEGREE-DAYS (°C)			WIND				PRECIPITATION					
	Relative (%)		Vapour Pressure kPa								RAINFALL (mm)		SNOWFALL (cm)			
	6 a.m. Average	3 p.m. Average		Heating (<18°)	Cooling (>18°)	Growing (>5°)	Prevailing Direction	% Frequency	Average Speed (km/h)	Extreme Gust (km/h)	Average	Wettest Day	Average	Snowiest Day	Average Snow on Ground	Total (mm)
Jan	73	71	0.21	991.8	0.0	0.1	NNE	12.9	19.9	161	8.1	28.7	93.4	48.3	94	95.5
Feb	72	70	0.22	861.1	0.0	0.0	N	11.1	18.6	130	8.7	88.6	74.0	41.9	114	79.5
Mar	74	72	0.31	763.3	0.0	0.1	N	12.6	19.7	121	14.4	39.6	70.0	50.8	74	82.8
Apr	76	70	0.46	540.6	0.0	1.6	E	14.3	18.3	119	45.7	74.9	33.0	43.4	16	78.4
May	76	66	0.66	376.3	0.0	52.0	E	16.4	17.7	121	77.5	69.6	6.0	29.2	0	84.0
Jun	78	68	1.02	188.6	1.2	202.9	E	16.6	15.7	129	90.2	68.1	T	0.5	0	90.2
Jul	83	71	1.34	91.0	6.2	318.2	E	14.8	14.2	103	97.0	84.8	0.0	0.0	0	97.0
Aug	85	70	1.26	123.8	2.9	282.1	E	12.2	14.4	113	104.1	76.5	0.0	0.0	0	104.1
Sep	86	70	0.93	262.8	0.3	130.5	E	11.5	15.5	154	112.2	98.6	T	T	0	112.2
Oct	83	69	0.65	446.2	0.0	25.6	W	10.8	16.4	122	85.3	53.1	10.6	28.2	1	96.5
Nov	81	75	0.45	614.5	0.0	2.5	W	10.0	18.0	130	49.4	114.6	50.8	45.4	18	100.1
Dec	75	73	0.25	894.0	0.0	0.0	NNE	14.1	19.3	159	18.1	42.9	89.1	52.1	60	104.6
Ann			0.65	6154.0	10.6	1015.6	E	11.2	17.3	161	710.7	114.6	426.9	52.1		1124.9

AVERAGE FROST DATA

Last frost (spring) May 27
First frost (fall) September 19
Frost-free period 114 days

CLIMATIC EXTREMES

Wettest Month on Record   252.0 mm   November 1950
Highest Temperature on Record   32.2°   July 10, 1955
Lowest Temperature on Record   -43.3°   January 21, 1950

STATION INFORMATION

SEPT-ÎLES   50°13'N   66°16'W   55 m

The airport is located on the north shore of the St. Lawrence River, 9 km east of the town of Sept-Îles. The region south of the airport is dominated by several bays along the St. Lawrence River. The terrain surrounding the airport is generally flat and is interspersed with numerous marshy areas. The highest elevation, 344 m is found 19 km to the northwest. Elevations ranging from 460 to 580 m can be found 25 to 35 km north of the airport.

Montréal, Québec

LONG-TERM AVERAGES AND EXTREMES OF CLIMATE DATA

| | TEMPERATURE (°C) AVERAGE | | | EXTREMES | | AVERAGE NUMBER OF DAYS WITH | | | | Freezing | | | | BRIGHT SUNSHINE | SUNRISE/SUNSET | |
	Midafternoon	Early Morning	Daily	Highest	Lowest	Winds (>63 km/h)	Hail	Thunder	Fog	Temp.	Precip.	Rain	Snow	(hours)	Local Standard Time (h:min) on the 15th of Month	
Jan	-5.7	-14.6	-10.2	13.9	-37.8	*	0	0	2	31	3	4	16	106.0	7:32	16:36
Feb	-4.4	-13.5	-9.0	12.2	-33.9	1	0	0	2	28	3	4	12	128.4	6:58	17:20
Mar	1.6	-6.7	-2.5	25.6	-29.4	*	*	*	2	28	2	6	9	155.2	6:09	17:59
Apr	10.6	0.8	5.7	28.4	-15.0	*	*	1	1	13	*	11	3	188.9	5:11	18:39
May	18.5	7.4	13.0	33.9	-4.4	*	*	2	1	1	0	12	*	241.5	4:26	19:17
Jun	23.6	12.9	18.3	35.0	1.1	*	*	5	1	0	0	12	0	249.0	4:06	19:45
Jul	26.1	15.6	20.9	35.6	7.0	0	*	7	1	0	*	12	0	274.5	4:20	19:42
Aug	24.8	14.3	19.6	37.8	3.3	*	*	6	1	0	0	12	0	239.6	4:54	19:05
Sep	19.9	9.6	14.8	32.8	-2.2	0	0	3	2	*	0	12	0	168.9	5:32	18:09
Oct	13.3	4.1	8.7	28.3	-7.2	*	*	1	2	7	*	12	1	136.6	6:10	17:12
Nov	5.4	-1.5	2.0	21.7	-19.4	*	*	*	3	18	1	11	6	85.8	6:53	16:26
Dec	-3.0	-10.8	-6.9	16.7	-32.4	*	0	*	2	29	4	6	15	79.6	7:27	16:12
Ann	10.9	1.5	6.2	37.8	-37.8	1	*	25	20	155	13	114	62	2054.0		

| | HUMIDITY Relative (%) | | Vapour Pressure kPa | DEGREE-DAYS (°C) | | | WIND | | | | PRECIPITATION RAINFALL (mm) | | SNOWFALL (cm) | | | Total (mm) |
	6 a.m. Average	3 p.m. Average		Heating (<18°)	Cooling (>18°)	Growing (>5°)	Prevailing Direction	% Frequency	Average Speed (km/h)	Extreme Gust (km/h)	Average	Wettest Day	Average	Snowiest Day	Average Snow on Ground	Total (mm)
Jan	77	70	0.26	874.0	0.0	0.0	WSW	19.3	18.3	117	23.7	31.0	52.7	32.8	25	72.0
Feb	77	68	0.27	762.5	0.0	0.0	WSW	16.7	17.9	109	14.9	31.5	53.6	39.4	28	65.2
Mar	77	64	0.40	636.5	0.0	4.0	WSW	14.4	17.9	161	36.8	32.0	35.7	43.2	6	73.6
Apr	77	55	0.62	368.5	0.2	64.1	W	11.9	16.9	106	63.5	34.5	9.7	25.7	0	74.1
May	76	51	0.98	165.8	10.5	249.0	SW	14.1	15.3	101	63.9	37.6	1.7	21.8	0	65.6
Jun	79	56	1.46	40.6	49.8	399.1	SW	16.9	14.5	111	82.2	54.1	0.0	0.0	0	82.2
Jul	83	55	1.72	8.3	98.7	493.4	SW	17.9	13.1	126	90.0	55.4	0.0	0.0	0	90.0
Aug	86	57	1.66	23.1	71.8	451.7	SW	14.6	12.2	105	91.9	68.8	0.0	0.0	0	91.1
Sep	87	60	1.30	116.1	18.9	292.9	SW	12.2	13.1	90	88.4	81.9	0.0	6.1	0	88.4
Oct	84	62	0.87	289.3	0.9	131.6	W	12.3	14.8	117	73.8	54.9	1.7	14.2	T	75.5
Nov	82	71	0.59	481.0	0.0	25.3	W	13.8	16.6	109	61.2	55.1	21.2	30.5	6	81.0
Dec	80	73	0.33	771.8	0.0	1.8	W	15.6	16.8	103	32.6	42.9	58.8	37.8	22	86.7
Ann			0.87	4537.5	250.8	2112.9	WSW	13.7	15.6	161	722.9	81.9	235.1	43.2		946.2

AVERAGE FROST DATA
Last frost (spring) May 3
First frost (fall) October 8
Frost-free period 157 days

CLIMATIC EXTREMES
Wettest Month on Record 227.6 mm September 1975
Highest Temperature on Record 37.6° August 1, 1975
Lowest Temperature on Record -37.8° January 15, 1957

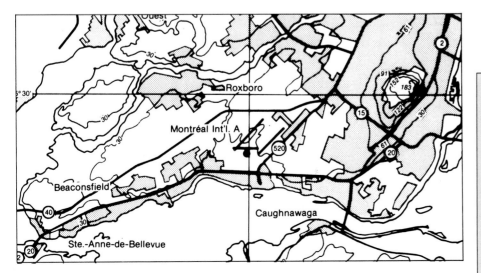

STATION INFORMATION

MONTRÉAL DORVAL 45°28'N 73°45'W 36 m
The area surrounding the airport is generally flat and level, with the exception of Mount Royal located about 11 km to the east-northeast with a maximum elevation of 233 m. The north branch of the St. Lawrence River flows eastward about 5 km north of the site, and the south branch of the river passes 3 km to the south. The site is about 13 km west-southwest of the centre of Montréal. Observations taken at the present site are considered representative of the suburban area surrounding the city of Montréal.

Toronto, Ontario

LONG-TERM AVERAGES AND EXTREMES OF CLIMATE DATA

| | TEMPERATURE (°C) | | | | | AVERAGE NUMBER OF DAYS WITH | | | | | | | | BRIGHT SUNSHINE | SUNRISE/SUNSET | |
| | AVERAGE | | | EXTREMES | | | | | | Freezing | | | | | | |
	Midafternoon	Early Morning	Daily	Highest	Lowest	Winds (>63 km/h)	Hail	Thunder	Fog	Temp.	Precip.	Rain	Snow	(hours)	Local Standard Time (h:min) on the 15th of Month	
Jan	-2.5	-10.9	-6.7	16.7	-31.1	*	*	0	2	30	3	4	12	92.1	7:49	17:05
Feb	-1.6	-10.5	-6.1	12.2	-31.1	*	0	*	3	27	2	4	10	111.6	7:18	17:46
Mar	3.3	-5.2	-1.0	25.6	-28.9	*	*	1	3	27	2	7	8	145.0	6:32	18:22
Apr	11.5	0.8	6.2	29.4	-17.2	*	*	2	3	14	*	10	2	182.3	5:36	18:59
May	18.4	6.1	12.3	34.4	-5.6	*	*	3	3	3	0	11	*	232.7	4:53	19:34
Jun	23.9	11.5	17.7	36.7	0.6	*	*	5	2	0	0	9	0	252.5	4:35	20:00
Jul	26.8	14.2	20.6	36.1	3.9	0	*	6	2	0	0	9	0	280.5	4:49	19:58
Aug	25.8	13.6	19.7	38.3	1.1	*	*	5	3	0	0	10	0	251.5	5:21	19:23
Sep	21.3	9.6	15.5	36.7	-3.9	*	*	3	3	1	0	9	0	191.8	5:56	18:31
Oct	14.6	3.9	9.3	30.6	-8.3	0	*	1	4	7	0	10	*	149.1	6:31	17:37
Nov	7.2	-0.6	3.3	25.0	-18.3	*	0	1	3	18	*	10	4	81.1	7:11	16:54
Dec	0.4	-7.4	-3.5	18.9	-31.1	*	0	*	4	28	3	6	11	75.2	7:44	16:41
Ann	12.4	2.1	7.3	38.3	-31.1	*	*	27	35	155	10	99	47	2045.4		

| | HUMIDITY | | | DEGREE-DAYS (°C) | | | WIND | | | | PRECIPITATION | | | | | |
| | Relative (%) | | Vapour Pressure kPa | | | | | | Average Speed (km/h) | Extreme Gust (km/h) | RAINFALL (mm) | | SNOWFALL (cm) | | | |
	6 a.m. Average	3 p.m. Average		Heating (<18°)	Cooling (>18°)	Growing (>5°)	Prevailing Direction	% Frequency			Average	Wettest Day	Average	Snowiest Day	Average Snow on Ground	Total (mm)
Jan	83	76	0.34	766.6	0.0	0.5	WSW	13.1	18.4	115	21.3	58.7	33.4	36.8	10	50.4
Feb	83	74	0.35	680.0	0.0	0.3	N	12.7	17.6	97	20.6	31.8	26.6	39.9	5	46.0
Mar	83	69	0.47	588.1	0.0	8.2	N	12.2	17.6	124	37.1	41.7	22.3	32.3	T	61.1
Apr	82	57	0.67	355.4	0.5	74.5	N	12.0	17.3	111	61.8	41.7	7.4	26.7	0	70.0
May	82	54	0.99	187.5	9.3	226.9	N	11.5	14.9	98	65.8	92.7	0.1	2.3	0	66.0
Jun	84	56	1.44	51.9	43.7	381.8	N	11.3	13.4	102	67.1	43.9	0.0	T	0	67.1
Jul	85	53	1.66	12.1	91.5	482.4	N	11.2	12.5	89	71.4	118.5	0.0	0.0	0	71.4
Aug	89	56	1.66	20.4	73.7	456.4	N	11.0	12.3	93	76.8	80.8	0.0	0.0	0	76.8
Sep	90	59	1.36	102.0	25.9	313.9	N	11.3	13.0	92	63.5	108.0	0.0	T	0	63.5
Oct	88	62	0.94	272.2	1.4	144.9	W	10.8	14.1	96	61.0	121.4	0.9	7.4	0	61.8
Nov	86	72	0.67	440.9	0.0	33.3	W	13.4	16.6	97	55.2	86.1	8.0	33.5	1	62.7
Dec	85	78	0.43	666.7	0.0	3.7	W	11.4	17.0	95	35.6	40.9	32.5	28.2	6	64.7
Ann			0.92	4143.8	246.0	2126.8	N	11.0	15.4	124	637.2	121.4	131.2	39.9		761.5

AVERAGE FROST DATA
Last frost (spring) May 8
First frost (fall) October 5
Frost-free period 149 days

CLIMATIC EXTREMES
Wettest Month on Record 213.9 mm October 1954
Highest Temperature on Record 38.3° August 25, 1948
Lowest Temperature on Record -31.3° January 4, 1981

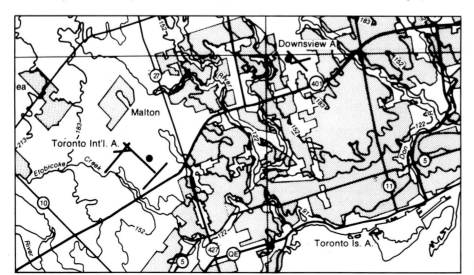

STATION INFORMATION

TORONTO PEARSON 43°40'N 79°38'W 173 m
The area surrounding the airport is slightly rolling to flat. There is a gentle rise inland from Lake Ontario (16 km to the south) to higher ground above the Niagara Escarpment about 24 km west to northwest of the airport. The site is about 19 km to the west-northwest of the Toronto City Hall. Most of the surrounding countryside is open, however, since 1953 urban growth is very much in evidence with numerous apartment towers, hotels and industrial complexes to the northeast, east and southeast.

Winnipeg, Manitoba
LONG-TERM AVERAGES AND EXTREMES OF CLIMATE DATA

| | TEMPERATURE (°C) | | | | | AVERAGE NUMBER OF DAYS WITH | | | | | | | | BRIGHT SUNSHINE | SUNRISE/SUNSET | |
| | AVERAGE | | | EXTREMES | | | | | | Freezing | | | | | | |
	Midafternoon	Early Morning	Daily	Highest	Lowest	Winds (> 63 km/h)	Hail	Thunder	Fog	Temp.	Precip.	Rain	Snow	(hours)	Local Standard Time (h:min) on the 15th of Month	
Jan	-14.3	-24.2	-19.3	7.8	-42.2	*	0	0	2	31	2	*	12	121.2	8:22	16:54
Feb	-10.1	-21.0	-15.6	11.7	-45.0	*	0	0	2	28	2	1	9	144.2	7:41	17:45
Mar	-2.8	-13.5	-8.2	23.3	-37.8	*	0	*	3	30	1	2	9	176.2	6:45	18:32
Apr	8.9	-2.2	3.4	34.3	-26.3	*	*	1	1	21	1	5	4	219.8	5:38	19:20
May	18.0	4.5	11.3	37.0	-11.1	*	1	3	1	8	*	10	1	265.6	4:44	20:06
Jun	23.1	10.5	16.8	36.7	-3.3	1	1	6	1	*	0	11	*	276.1	4:19	20:39
Jul	25.9	13.3	19.6	37.8	1.1	*	1	8	1	0	0	11	0	315.6	4:36	20:34
Aug	24.7	11.8	18.3	40.6	0.6	*	*	6	1	0	0	11	0	283.3	5:17	19:50
Sep	18.4	6.3	12.4	36.2	-7.2	*	*	3	2	3	0	11	*	184.6	6:03	18:46
Oct	11.5	0.7	6.1	29.4	-17.2	*	0	*	2	15	*	7	2	151.5	6:49	17:41
Nov	-0.3	-8.8	-4.5	23.9	-32.2	*	*	*	2	28	3	2	9	90.7	7:40	16:47
Dec	-9.4	-18.6	-14.0	11.7	-37.8	*	0	*	2	31	3	1	11	92.6	8:20	16:28
Ann	7.8	-3.4	2.2	40.6	-45.0	1	3	27	20	195	12	72	57	2321.4		

| | HUMIDITY | | | DEGREE-DAYS (°C) | | | WIND | | | | PRECIPITATION | | | | | |
| | Relative (%) | | | | | | | | | | RAINFALL (mm) | | SNOWFALL (cm) | | | |
	6 a.m. Average	3 p.m. Average	Vapour Pressure kPa	Heating (<18°)	Cooling (>18°)	Growing (>5°)	Prevailing Direction	% Frequency	Average Speed (km/h)	Extreme Gust (km/h)	Average	Wettest Day	Average	Snowiest Day	Average Snow on Ground	Total (mm)
Jan	81	76	0.14	1154.7	0.0	0.0	S	15.3	18.6	98	0.2	3.8	23.7	19.1	32	21.3
Feb	82	77	0.19	948.6	0.0	0.0	S	17.1	18.1	129	0.7	7.6	18.9	23.6	29	17.5
Mar	83	73	0.32	811.3	0.0	1.6	S	15.4	19.3	113	3.3	30.0	21.1	35.6	13	22.7
Apr	82	57	0.54	439.4	1.0	50.0	S	12.2	20.9	106	27.1	33.3	11.3	21.3	1	38.5
May	77	46	0.81	219.3	10.1	203.8	S	10.7	20.2	109	63.2	60.2	2.5	21.1	0	65.7
Jun	80	50	1.25	69.9	34.2	354.3	S	13.1	18.1	127	80.1	60.7	T	0.3	0	80.1
Jul	85	52	1.56	21.1	71.9	453.8	S	13.4	16.0	127	75.9	69.1	0.0	0.0	0	75.9
Aug	87	51	1.44	43.3	52.0	411.7	S	12.9	16.4	122	75.2	83.8	0.0	0.0	0	75.2
Sep	86	53	1.02	178.1	8.7	223.8	S	15.2	18.5	98	53.0	65.0	0.2	1.8	0	53.3
Oct	84	57	0.69	369.6	0.5	82.1	S	15.7	19.6	102	25.9	74.4	5.2	24.6	1	30.9
Nov	84	73	0.39	676.0	0.0	3.8	S	15.1	19.4	124	5.5	17.0	21.9	27.7	11	25.2
Dec	82	78	0.21	991.8	0.0	0.0	S	18.0	18.6	89	0.9	5.8	20.7	21.6	18	19.2
Ann			0.71	5923.1	178.4	1784.9	S	14.5	18.6	129	411.0	83.8	125.5	35.6		525.5

AVERAGE FROST DATA
Last frost (spring) May 23
First frost (fall) September 22
Frost-free period 121 days

CLIMATIC EXTREMES
Wettest Month on Record   197.4 mm   July 1953
Highest Temperature on Record   40.6°   August 7, 1949
Lowest Temperature on Record   -45.0°   February 18, 1966

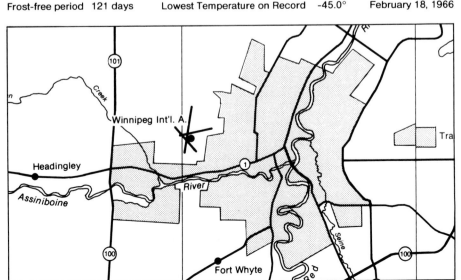

STATION INFORMATION
WINNIPEG INT'L   49°54'N   97°14'W   239 m
The Winnipeg urban area extends 8 to 12 km beyond the airport from the northeast through north and west-southwest directions. The airport is situated in the broad flat valley of the Red River. The surrounding Prairie farmland varies little in elevation except at the nearby rivers where the elevation drops to about 230m.

Churchill, Manitoba

LONG-TERM AVERAGES AND EXTREMES OF CLIMATE DATA

	TEMPERATURE (°C)					AVERAGE NUMBER OF DAYS WITH								BRIGHT SUNSHINE	SUNRISE/SUNSET	
	AVERAGE			EXTREMES						Freezing					Local Standard Time (h:min) on the 15th of Month	
	Midafternoon	Early Morning	Daily	Highest	Lowest	Winds (>63 km/h)	Hail	Thunder	Fog	Temp.	Precip.	Rain	Snow	(hours)		
Jan	-23.6	-31.4	-27.5	0.0	-45.0	1	0	0	2	31	*	*	11	80.4	8:56	15:55
Feb	-21.7	-30.0	25.9	1.1	-45.4	1	0	0	2	28	*	0	10	131.7	7:52	17:09
Mar	-15.7	-25.0	-20.4	5.6	-43.9	1	0	0	1	31	1	*	10	188.6	6:35	18:16
Apr	-5.4	-14.8	-10.1	28.2	-33.3	*	0	0	4	30	2	1	10	203.7	5:06	19:27
May	-2.2	-5.1	-1.5	27.2	-21.7	1	*	*	7	29	5	5	7	195.3	3:49	20:36
Jun	10.8	1.5	6.2	31.1	-9.4	*	*	2	8	13	2	9	2	233.7	3:06	21:27
Jul	16.8	6.8	11.8	33.9	-2.2	*	*	2	7	1	*	11	0	285.1	3:29	21:15
Aug	15.3	7.2	11.3	32.8	-2.2	1	*	2	6	*	0	13	0	232.1	4:34	20:08
Sep	8.5	2.2	5.4	27.8	-11.7	1	0	1	4	9	1	12	4	110.8	5:43	18:40
Oct	1.3	-4.4	-1.5	20.6	-24.4	2	0	0	4	25	3	6	14	61.7	6:51	17:14
Nov	-8.3	-15.9	-12.1	7.2	-36.1	2	0	0	2	30	4	1	18	49.5	8:05	15:57
Dec	-18.2	-26.1	-22.2	2.2	-40.0	1	0	0	1	31	1	*	14	55.3	9:01	15:21
Ann	-3.2	-11.3	-7.2	33.9	-45.4	11	*	7	48	258	19	58	100	1827.9		

	HUMIDITY			DEGREE-DAYS (°C)			WIND				PRECIPITATION					
	Relative (%)		Vapour Pressure kPa						Average Speed (km/h)	Extreme Gust (km/h)	RAINFALL (mm)		SNOWFALL (cm)			Total (mm)
	6 a.m. Average	3 p.m. Average		Heating (<18°)	Cooling (>18°)	Growing (>5°)	Prevailing Direction	% Frequency			Average	Wettest Day	Average	Snowiest Day	Average Snow on Ground	
Jan	70	70	0.06	1410.4	0.0	0.0	W	21.3	24.2	126	T	0.3	16.9	16.0	40	15.3
Feb	71	71	0.07	1240.2	0.0	0.0	WNW	27.2	24.1	105	0.1	1.3	14.6	12.7	46	13.1
Mar	74	73	0.12	1191.4	0.0	0.0	WNW	19.3	22.2	105	0.6	15.2	18.6	22.6	50	18.1
Apr	82	78	0.26	843.0	0.0	1.4	NW	11.0	22.6	97	2.0	8.4	22.3	25.4	26	22.9
May	86	79	0.47	602.9	0.0	10.1	NNW	12.0	22.2	108	13.5	22.4	19.5	47.6	2	31.9
Jun	86	72	0.75	356.4	1.8	78.8	N	9.4	20.7	109	39.9	32.5	3.5	15.7	0	43.5
Jul	86	67	1.05	200.1	8.9	215.3	N	9.2	19.3	97	45.6	52.3	0.0	T	0	45.6
Aug	86	70	1.05	212.3	3.6	194.6	NNW	9.8	20.5	124	58.3	51.1	0.0	T	0	58.3
Sep	86	73	0.74	378.2	0.0	50.8	NNW	11.5	23.7	151	44.5	42.2	6.4	17.5	1	50.9
Oct	86	80	0.49	605.6	0.0	3.6	NW	12.4	24.9	121	15.4	26.2	29.3	36.1	6	43.0
Nov	83	81	0.24	903.7	0.0	0.0	WNW	15.3	25.6	122	1.0	4.0	41.6	35.1	25	38.8
Dec	74	74	0.10	1245.7	0.0	0.0	W	17.3	22.7	101	0.2	1.8	22.8	21.8	34	20.9
Ann			0.45	9189.9	14.3	554.6	WNW	12.7	22.7	151	221.1	52.3	195.5	47.6		402.3

AVERAGE FROST DATA
Last frost (spring) June 24
First frost (fall) September 9
Frost-free period 76 days

CLIMATIC EXTREMES
Wettest Month on Record 135.1 mm June 1960
Highest Temperature on Record 33.9° July 4, 1975
Lowest Temperature on Record -45.4° February 13, 1979

STATION INFORMATION
CHURCHILL 58°45'N 94°04'W 29 m
The airport is located 6 km east-southeast of the town of Churchill and 1 km south of the settlement of Fort Churchill in northeastern Manitoba. Hudson Bay is only 2 km north of the station. The Churchill River, is 4 km to the west. The surrounding countryside is flat with numerous lakes, swamp and marsh. Rarely does the elevation exceed 30 m. There are a number of eskers in the immediate area of the station; two with lengths greater than 5 km are located 19 km to the east and 12 km to the southwest. The vegetation of the area comprises tundra, especially along the southern shore of Hudson Bay and the northern reaches of the Churchill River, and sporadic tree growth.

Regina, Saskatchewan
LONG-TERM AVERAGES AND EXTREMES OF CLIMATE DATA

| | TEMPERATURE (°C) | | | | | AVERAGE NUMBER OF DAYS WITH | | | | | | | | BRIGHT SUNSHINE | SUNRISE/SUNSET |
| | AVERAGE | | | EXTREMES | | | | | | Freezing | | | | | |
	Midafternoon	Early Morning	Daily	Highest	Lowest	Winds (>63 km/h)	Hail	Thunder	Fog	Temp.	Precip.	Rain	Snow	(hours)	Local Standard Time (h:min) on the 15th of Month
Jan	-12.6	-23.2	-17.9	8.9	-50.0	1	0	0	3	31	2	*	13	99.9	8:54 17:22
Feb	-8.1	-19.1	-13.6	15.6	-47.8	1	0	0	5	28	3	*	10	121.3	8:12 18:14
Mar	-2.3	-13.3	-7.8	24.4	-40.6	*	0	0	5	30	2	1	.9	156.4	7:14 19:01
Apr	9.4	-2.8	3.3	32.8	-28.9	1	*	*	2	22	1	4	4	209.3	6:07 19:51
May	18.2	3.9	11.1	37.2	-13.3	1	*	3	1	7	*	9	1	277.5	5:12 20:38
Jun	22.7	9.0	15.9	39.4	-5.6	1	1	6	1	1	0	12	0	283.0	4:47 21:11
Jul	26.1	11.7	18.9	43.3	-2.2	*	*	7	1	0	0	10	0	342.2	5:03 21:06
Aug	25.2	10.4	17.8	40.6	-5.0	*	*	5	1	*	0	9	0	294.7	5:46 20:21
Sep	18.6	4.6	11.7	37.2	-16.1	1	*	2	1	5	*	8	1	190.8	6:33 19:16
Oct	11.9	-1.5	5.2	31.1	-26.1	1	0	*	2	20	*	4	2	168.1	7:20 18:10
Nov	0.2	-10.4	-5.1	23.3	-37.2	1	0	0	3	29	3	1	8	104.1	8:11 17:15
Dec	-7.7	-18.0	-12.8	15.0	-48.3	1	0	0	4	31	3	1	10	83.9	8:52 16:55
Ann	8.5	-4.1	2.2	43.3	-50.0	9	1	23	29	204	14	59	58	2331.1	

| | HUMIDITY | | | DEGREE-DAYS (°C) | | | WIND | | | | PRECIPITATION | | | | | |
| | Relative (%) | | | | | | | | | | RAINFALL (mm) | | SNOWFALL (cm) | | | |
	6 a.m. Average	3 p.m. Average	Vapour Pressure kPa	Heating (<18°)	Cooling (>18°)	Growing (>5°)	Prevailing Direction	% Frequency	Average Speed (km/h)	Extreme Gust (km/h)	Average	Wettest Day	Average	Snowiest Day	Average Snow on Ground	Total (mm)
Jan	79	77	0.16	1113.9	0.0	0.0	SE	16.4	21.8	122	0.3	4.3	20.0	14.0	25	16.6
Feb	82	78	0.22	894.2	0.0	0.0	SE	19.0	21.6	122	0.8	7.1	18.3	19.1	22	16.1
Mar	85	73	0.33	800.1	0.0	1.7	SE	18.0	22.0	95	2.1	17.8	18.3	25.4	10	17.8
Apr	83	54	0.53	440.1	0.2	45.2	SE	14.3	22.6	126	14.3	30.2	10.9	23.2	1	23.7
May	78	44	0.78	221.0	5.7	195.9	SE	12.4	22.1	135	43.5	57.2	3.2	19.8	0	46.4
Jun	79	46	1.13	84.0	20.5	326.5	SE	9.9	19.9	153	79.6	160.3	0.0	7.6	0	79.6
Jul	82	45	1.37	28.9	57.6	431.6	SE	11.2	17.7	137	53.3	76.5	0.0	T	0	53.3
Aug	81	43	1.23	50.7	45.9	398.2	SE	13.0	18.1	137	44.8	78.7	0.0	0.0	0	44.8
Sep	83	47	0.87	196.9	6.5	205.7	SE	12.2	20.4	117	35.0	79.8	1.8	21.6	0	36.7
Oct	83	51	0.59	397.9	0.1	69.3	SE	13.0	20.5	127	11.4	31.2	8.2	21.3	1	18.8
Nov	84	69	0.36	693.0	0.0	3.1	SE	14.5	20.9	132	1.4	20.8	14.2	23.9	7	13.5
Dec	82	78	0.23	955.9	0.0	0.0	SE	18.3	21.7	105	0.7	9.7	20.8	14.2	16	16.7
Ann			0.65	5876.6	136.5	1677.2	SE	14.4	20.8	153	287.2	160.3	115.7	25.4		384.0

AVERAGE FROST DATA
Last frost (spring) May 24
First frost (fall) September 11
Frost-free period 109 days

CLIMATIC EXTREMES
Wettest Month on Record   207.0 mm   July 1901
Highest Temperature on Record   43.3°   July 5, 1937
Lowest Temperature on Record   -50.0°   January 1, 1885

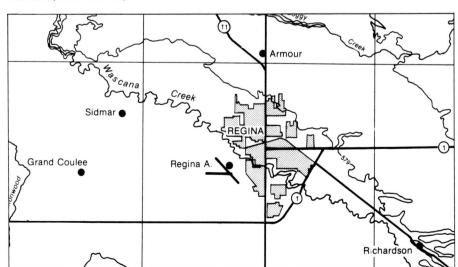

STATION INFORMATION
REGINA   50°27'N   104°37'W   577 m
The area surrounding Regina is a level plain. The city is situated on the banks of Wascana Creek, which lies in a very shallow basin running southeast to northwest. The land rises slowly to the northeast and reaches a maximum elevation of about 700 m, approximately 35 km northeast of the airport. The area within a 32 km radius of the site is devoted to agriculture, the main crop being wheat.

Edmonton, Alberta

LONG-TERM AVERAGES AND EXTREMES OF CLIMATE DATA

	TEMPERATURE (°C)					AVERAGE NUMBER OF DAYS WITH									BRIGHT SUNSHINE	SUNRISE/SUNSET	
	AVERAGE			EXTREMES						Freezing							
	Midafternoon	Early Morning	Daily	Highest	Lowest	Winds (>63 km/h)	Hail	Thunder	Fog	Temp.	Precip.	Rain	Snow	(hours)	Local Standard Time (h:min) on the 15th of Month		
Jan	-10.7	-19.2	-15.0	11.7	-44.4	*	*	0	3	31	1	1	13	90.0	8:44	16:43
Feb	-4.8	-14.3	-9.6	13.9	-46.1	*	0	0	2	28	1	1	9	116.3	7:55	17:42
Mar	-0.3	-9.7	-5.0	20.6	-36.1	0	0	0	2	29	1	1	10	167.5	6:51	18:36
Apr	9.6	-1.2	4.2	31.1	-25.6	0	0	*	1	19	1	3	5	228.3	5:36	19:32
May	17.2	5.4	11.3	30.6	-12.2	*	*	2	*	2	*	9	1	277.6	4:35	20:26
Jun	20.7	9.5	15.1	34.4	-1.1	*	1	5	1	0	*	13	*	271.7	4:04	21:04
Jul	23.0	11.8	17.4	34.4	0.6	0	1	8	*	0	0	13	0	306.4	4:22	20:57
Aug	21.6	10.7	16.2	33.3	0.6	*	1	6	1	0	0	13	0	276.8	5:12	20:06
Sep	16.5	5.6	11.0	33.9	-11.7	*	*	1	1	3	*	9	1	182.2	6:06	18:53
Oct	11.1	0.4	5.8	28.3	-25.0	*	*	*	1	14	*	4	2	161.8	6:59	17:41
Nov	0.4	-7.8	-3.7	21.7	-32.8	*	*	0	3	28	2	2	7	107.2	7:58	16:39
Dec	-6.2	-14.5	-10.4	16.1	-48.3	*	0	0	2	31	2	1	11	77.9	8:44	16:14
Ann	8.2	-1.9	3.1	34.4	-48.3	*	3	22	17	185	8	70	59	2263.7		

	HUMIDITY			DEGREE-DAYS (°C)			WIND				PRECIPITATION					
	Relative (%)		Vapour Pressure kPa								RAINFALL (mm)		SNOWFALL (cm)			Total (mm)
	6 a.m. Average	3 p.m. Average		Heating (<18°)	Cooling (>18°)	Growing (>5°)	Prevailing Direction	% Frequency	Average Speed (km/h)	Extreme Gust (km/h)	Average	Wettest Day	Average	Snowiest Day	Average Snow on Ground	
Jan	78	72	0.19	1021.5	0.0	0.1	S	12.4	13.0	114	1.4	9.4	27.2	13.7	23	24.6
Feb	78	68	0.26	779.7	0.0	0.4	S	14.4	13.0	105	0.5	4.1	21.3	18.0	22	18.8
Mar	80	63	0.34	713.7	0.0	2.6	S	14.7	13.8	93	1.6	6.6	18.7	18.0	7	18.5
Apr	76	47	0.49	413.4	0.3	50.5	S	13.8	15.5	90	8.8	27.2	13.2	38.1	0	21.7
May	70	40	0.70	209.1	2.8	201.9	S	10.2	16.1	103	39.4	43.9	3.1	13.5	0	42.5
Jun	76	47	1.01	98.7	12.2	303.6	WNW	10.9	15.4	117	77.3	80.0	T	1.0	0	77.3
Jul	81	51	1.27	48.7	31.5	385.7	WNW	11.2	14.0	108	88.7	114.0	0.0	0.0	0	88.7
Aug	85	52	1.23	78.0	21.1	346.1	S	10.4	13.4	109	77.9	59.7	0.0	T	0	77.9
Sep	84	53	0.89	210.4	1.7	189.2	S	11.0	14.6	117	37.0	51.3	2.2	7.4	T	39.1
Oct	77	50	0.58	379.9	0.1	75.9	S	15.9	14.3	113	9.1	18.0	7.5	21.6	1	16.6
Nov	78	65	0.36	650.6	0.0	4.2	S	15.5	13.5	100	2.5	8.9	15.4	39.9	6	15.7
Dec	78	71	0.25	880.0	0.0	0.2	S	13.3	13.0	98	1.4	5.8	27.1	25.8	15	24.7
Ann			0.63	5483.0	69.7	1560.4	S	12.3	14.1	117	345.6	114.0	135.7	39.9		466.1

AVERAGE FROST DATA
Last frost (spring) May 6
First frost (fall) September 24
Frost-free period 140 days

CLIMATIC EXTREMES
Wettest Month on Record   190.8 mm   July 1953
Highest Temperature on Record   34.4°   June 3, 1970
Lowest Temperature on Record   -48.3°   December 28, 1938

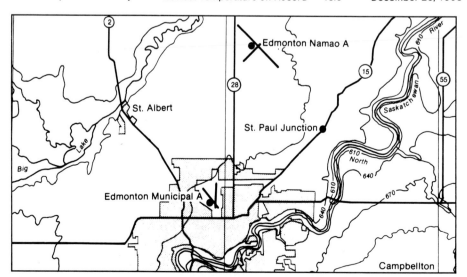

STATION INFORMATION

EDMONTON MUNICIPAL   53°34'N 113°31'W   671m
The airport is in an urban setting 3 km northwest of the main business district, about 30 km northwest of Edmonton Int'l A. The city occupies a rolling plain. The North Saskatchewan River flows northeast, 4 km south of the airport. Residential-commercial buildings surround the airport within a 10-km radius. Beyond this area the land use is mainly mixed farming.

Penticton, British Columbia

LONG-TERM AVERAGES AND EXTREMES OF CLIMATE DATA

	TEMPERATURE (°C) AVERAGE			TEMPERATURE (°C) EXTREMES		AVERAGE NUMBER OF DAYS WITH								BRIGHT SUNSHINE	SUNRISE/SUNSET	
	Midafternoon	Early Morning	Daily	Highest	Lowest	Winds (>63 km/h)	Hail	Thunder	Fog	Freezing Temp.	Freezing Precip.	Rain	Snow	(hours)	Local Standard Time (h:min) on the 15th of Month Sunrise	Sunset
Jan	-0.1	-5.3	-2.7	13.3	-26.7	*	0	0	*	26	1	4	11	48.0	7:50	16:25
Feb	4.1	-3.0	0.6	15.6	-26.7	*	*	0	1	21	*	5	5	75.2	7:10	17:16
Mar	9.1	-1.4	3.9	21.7	-17.8	*	*	*	*	21	0	6	2	140.2	6:14	18:01
Apr	15.4	1.8	8.6	28.9	-7.2	*	*	1	0	11	0	6	*	211.2	5:08	18:49
May	20.6	6.1	13.4	32.2	-5.6	*	*	1	0	2	0	9	0	246.1	4:16	19:34
Jun	24.6	9.8	17.2	36.7	0.0	0	*	2	0	*	0	8	0	262.6	3:51	20:06
Jul	28.6	12.0	20.3	40.6	2.2	0	*	4	0	0	0	6	0	311.3	4:07	20:01
Aug	27.4	11.6	19.5	38.9	2.9	0	*	3	0	0	0	7	*	270.7	4:48	19:18
Sep	21.8	7.6	14.7	34.4	-2.8	0	0	1	0	1	0	7	0	211.2	5:33	18:15
Oct	14.5	3.0	8.7	28.9	-7.8	*	*	*	*	9	0	7	*	157.1	6:18	17:11
Nov	6.3	-0.4	3.0	19.4	-18.9	*	0	*	*	16	*	7	3	59.8	7:08	16:18
Dec	2.1	-2.9	-0.4	14.4	-27.2	*	*	0	*	22	*	6	8	38.8	7:47	15:59
Ann	14.5	3.2	8.9	40.6	-27.2	*	*	12	1	129	1	78	29	2032.2		

	HUMIDITY Relative (%)		Vapour Pressure kPa	DEGREE-DAYS (°C)			WIND				PRECIPITATION RAINFALL (mm)		SNOWFALL (cm)			
	6 a.m. Average	3 p.m. Average		Heating (<18°)	Cooling (>18°)	Growing (>5°)	Prevailing Direction	% Frequency	Average Speed (km/h)	Extreme Gust (km/h)	Average	Wettest Day	Average	Snowiest Day	Average Snow on Ground	Total (mm)
Jan	79	74	0.43	641.6	0.0	2.0	S	29.0	16.5	82	8.2	15.7	29.0	19.6	6	32.0
Feb	81	67	0.50	493.5	0.0	4.3	S	25.4	14.3	93	9.9	18.5	11.4	23.1	2	19.8
Mar	78	50	0.52	438.1	0.0	25.9	S	22.0	13.1	113	13.1	16.3	4.4	10.0	0	17.3
Apr	79	41	0.63	282.1	0.0	111.8	N	19.1	11.6	121	21.0	27.4	0.2	4.1	0	21.4
May	72	41	0.86	147.7	3.4	258.7	N	22.3	10.5	97	29.1	29.7	T	·T	0	29.1
Jun	68	40	1.07	52.3	29.3	367.1	N	23.2	10.3	77	27.6	33.5	0.0	T	0	27.6
Jul	65	38	1.20	14.2	84.6	473.5	N	28.7	10.2	85	21.1	37.8	0.0	0.0	0	21.1
Aug	71	39	1.21	19.1	66.2	450.2	N	28.2	9.7	80	26.5	23.6	0.0	0.0	0	26.5
Sep	79	45	1.02	104.5	6.8	292.4	N	25.6	9.8	97	17.7	28.1	0.0	T	0	17.7
Oct	80	55	0.78	287.3	0.0	123.2	S	18.7	11.9	101	15.2	44.5	0.2	2.0	0	15.3
Nov	79	68	0.59	450.6	0.0	23.4	S	30.7	16.0	97	16.8	19.3	7.7	20.6	2	23.9
Dec	80	75	0.49	570.6	0.0	3.2	S	33.9	18.0	113	11.5	12.4	23.1	22.1	5	31.2
Ann			0.78	3501.6	190.3	2135.7	S	18.5	12.7	121	217.7	44.5	76.0	23.1		282.9

AVERAGE FROST DATA
Last frost (spring) May 8
First frost (fall) October 4
Frost-free period 148 days

CLIMATIC EXTREMES
Wettest Month on Record 86.1 mm August 1976
Highest Temperature on Record 40.6° July 17, 1941
Lowest Temperature on Record -27.2° December 30, 1968

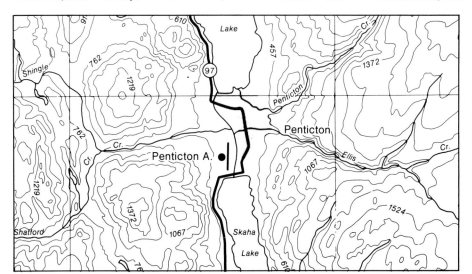

STATION INFORMATION
PENTICTON 49°28'N 119°36'W 344 m
The airport lies immediately south of the city of Penticton and 52 km north of the international border in the southern interior of British Columbia. The Okanagan Valley, in which the station is situated, is fairly flat and narrow and is bounded by the Cascade Mountains rising abruptly to the west and the Monashee Range to the east. Mountains in both of these ranges rise to heights over 2135 m. The southern shore of Lake Okanagan is 4 km north of the airport, from which point the lake extends 105 km northward.

Vancouver, British Columbia

LONG-TERM AVERAGES AND EXTREMES OF CLIMATE DATA

	TEMPERATURE (°C)					AVERAGE NUMBER OF DAYS WITH								BRIGHT SUNSHINE	SUNRISE/SUNSET
	AVERAGE			EXTREMES						Freezing						
	Midafternoon	Early Morning	Daily	Highest	Lowest	Winds (> 63 km/h)	Hail	Thunder	Fog	Temp.	Precip.	Rain	Snow	(hours)	Local Standard Time (h:min) on the 15th of Month	
Jan	5.2	-0.2	2.5	14.4	-17.8	*	0	*	6	15	1	17	6	53.5	8:03	16:41
Feb	7.8	1.4	4.6	15.0	-16.1	*	*	*	5	10	*	15	2	87.2	7:23	17:31
Mar	9.4	2.1	5,8	19.4	-9.4	*	*	*	2	9	*	15	2	129.3	6:28	18:16
Apr	12.8	4.7	8.8	24.4	-3.3	*	*	1	1	1	0	13	*	180.5	5:23	19:03
May	16.5	7.9	12.2	28.9	0.6	0	*	1	*	0	0	10	0	246.1	4:31	19:47
Jun	19.2	10.9	15.1	30.6	3.9	0	0	1	*	0	0	10	0	238.4	4:07	20:19
Jul	21.9	12.6	17.3	31.7	6.7	0	0	1	1	0	0	6	0	307.1	4:23	20:14
Aug	21.5	12.6	17.1	33.3	6.1	0	0	1	2	0	*	8	0	256.2	5:03	19:31
Sep	18.3	10.1	14.2	28.9	0.0	*	*	1	7	0	0	10	0	183.1	5:48	18:29
Oct	13.6	6.4	10.0	23.5	-3.3	*	0	*	9	1	0	15	*	121.0	6:32	17:25
Nov	9.0	2.8	5.9	18.4	-12.2	*	*	*	6	8	*	18	1	69.3	7:21	16:33
Dec	6.5	1.2	3.9	14.9	-17.8	*	*	*	6	11	*	19	4	47.9	8:00	16:15
Ann	13.5	6.0	9.8	33.3	-17.8	*	*	6	45	55	1	156	15	1919.6		

	HUMIDITY			DEGREE-DAYS (°C)			WIND				PRECIPITATION					
	Relative (%)		Vapour Pressure kPa								RAINFALL (mm)		SNOWFALL (cm)			
	6 a.m. Average	3 p.m. Average		Heating (<18°)	Cooling (>18°)	Growing (>5°)	Prevailing Direction	% Frequency	Average Speed (km/h)	Extreme Gust (km/h)	Average	Wettest Day	Average	Snowiest Day	Average Snow on Ground	Total (mm)
Jan	88	81	0.66	479.7	0.0	15.6	E	31.3	12.2	97	130.7	68.3	25.7	29.7	2	153.8
Feb	89	77	0.73	379.0	0.0	24.9	E	26.0	12.4	119	107.1	43.9	7.5	18.3	1	114.7
Mar	87	70	0.74	378.9	0.0	44.6	E	22.6	13.5	108	95.1	49.3	6.6	25.9	0	101.0
Apr	85	65	0.85	277.2	0.0	113.2	E	18.9	13.3	100	59.3	44.5	0.3	3.6	0	59.6
May	84	62	1.05	180.0	0.4	223.4	E	17.4	11.8	72	51.6	28.8	0.0	T	0	51.6
Jun	84	64	1.27	91.5	3.8	302.3	E	18.6	11.5	69	45.2	40.4	0.0	0.0	0	45.2
Jul	85	62	1.45	39.4	17.9	381.5	E	17.7	11.4	71	32.0	45.2	0.0	0.0	0	32.0
Aug	89	64	1.49	41.5	13.3	374.8	E	20.4	10.6	85	41.1	31.8	0.0	0.0	0	41.1
Sep	92	70	1.33	114.7	0.9	276.2	E	18.2	10.6	82	67.1	49.5	0.0	0.0	0	67.1
Oct	92	78	1.07	248.0	0.0	155.9	E	23.2	11.2	126	114.0	60.7	T	0.3	0	114.0
Nov	89	80	0.82	362.8	0.0	53.5	E	27.3	12.2	129	147.0	61.0	22.1	22.1	T	150.1
Dec	89	83	0.73	438.0	0.0	27.8	E	32.5	13.0	100	165.2	89.4	31.2	31.2	4	182.4
Ann			1.02	3030.7	36.3	1993.7	E	22.9	12.0	129	1055.4	60.4	60.4	31.2		1112.6

AVERAGE FROST DATA

Last frost (spring)	March 31
First frost (fall)	November 3
Frost-free period	216 days

CLIMATIC EXTREMES

Wettest Month on Record	350.8 mm	November 1983
Highest Temperature on Record	33.3°	August 9, 1960
Lowest Temperature on Record	-17.8°	December 29, 1968

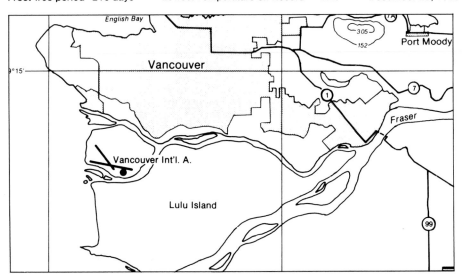

STATION INFORMATION

VANCOUVER INT'L 49°11'N 123°10'W 3 m

The airport is located on Sea Island in the Fraser River estuary. The island is about 5 km in diameter and its terrain is quite flat. About 6 km north of the airport the terrain rises 60 to 120 m in a ridge that runs from west to east through Vancouver. Northward from this ridge the terrain drops to sea level at Burrard Inlet. The Pacific Range of the Coast Mountains, 1200 to 1830 m in elevation, rises sharply from the north shore of Burrard Inlet about 25 km to the north and northeast of the airport. The flat farmland of the Fraser Lowlands extends to the south and east. The airport is about 16 km south of downtown Vancouver.

Prince Rupert, British Columbia

LONG-TERM AVERAGES AND EXTREMES OF CLIMATE DATA

	TEMPERATURE (°C)					AVERAGE NUMBER OF DAYS WITH								BRIGHT SUNSHINE	SUNRISE/SUNSET
	AVERAGE			EXTREMES						Freezing					
	Midafternoon	Early Morning	Daily	Highest	Lowest	Winds (>63 km/h)	Hail	Thunder	Fog	Temp.	Precip.	Rain	Snow	(hours)	Local Standard Time (h:min) on the 15th of Month
Jan	3.3	-3.7	-0.2	12.8	-24.4	1	1	0	2	22	0	15	9	48.0	8:55 16:47
Feb	5.7	-1.3	2.3	17.8	-17.2	1	1	0	1	16	0	17	5	63.3	8:05 17:48
Mar	6.5	-0.5	3.0	17.2	-17.2	*	1	0	1	18	0	19	8	93.9	6:59 18:43
Apr	9.2	1.5	5.4	23.4	-6.7	*	1	0	1	10	0	18	4	135.0	5:42 19:42
May	12.2	4.4	8.3	27.2	-1.7	*	*	0	1	3	0	18	0	189.2	4:39 20:37
Jun	14.1	7.4	10.8	25.5	1.1	0	0	0	4	0	0	17	0	150.7	4:07 21:17
Jul	16.0	9.6	12.8	27.8	2.8	*	0	0	6	0	0	16	0	142.7	4:26 21:09
Aug	16.5	9.6	13.1	28.7	2.8	0	0	0	8	0	0	16	0	138.3	5:17 20:16
Sep	15.1	7.4	11.4	23.9	-2.2	0	*	0	7	1	0	18	0	116.6	6:13 19:02
Oct	11.2	4.6	7.9	21.1	-5.6	*	1	0	2	5	0	24	0	64.8	7:08 17:47
Nov	7.1	0.3	3.8	18.9	-17.8	1	1	1	2	14	0	20	2	49.6	8:09 16:43
Dec	4.8	-1.6	1.6	18.9	-22.8	1	2	1	2	18	0	20	7	32.0	8:56 16:18
Ann	10.1	3.1	6.7	28.7	-24.4	4	8	2	37	107	0	218	35	1224.1	

	HUMIDITY			DEGREE-DAYS (°C)			WIND				PRECIPITATION					
	Relative (%)		Vapour Pressure kPa								RAINFALL (mm)		SNOWFALL (cm)			
	6 a.m. Average	3 p.m. Average		Heating (<18°)	Cooling (>18°)	Growing (>5°)	Prevailing Direction	% Frequency	Average Speed (km/h)	Extreme Gust (km/h)	Average	Wettest Day	Average	Snowiest Day	Average Snow on Ground	Total (mm)
Jan	86	78	0.54	560.2	0.0	6.7	SE	20.0	15.0	109	183.7	68.3	49.9	39.9	6	227.5
Feb	85	74	0.62	443.1	0.0	11.3	SE	28.4	16.9	113	191.6	95.0	23.2	23.9	3	222.1
Mar	87	70	0.63	465.2	0.0	9.0	SE	22.1	15.9	113	173.9	51.9	25.9	19.8	T	200.8
Apr	89	69	0.72	376.1	0.0	28.9	SE	20.6	16.5	109	180.9	98.6	7.3	10.4	0	190.0
May	90	71	0.87	299.5	0.0	100.7	SE	14.3	14.6	93	139.5	56.8	0.1	1.5	0	139.5
Jun	92	77	1.09	214.6	0.3	173.5	W	17.6	12.9	72	129.5	53.3	0.0	T	0	129.5
Jul	94	79	1.27	158.6	0.1	242.8	W	15.0	11.1	64	103.0	52.6	0.0	0.0	0	103.0
Aug	95	80	1.33	151.0	0.1	250.5	SE	12.5	11.1	64	158.4	87.6	0.0	0.0	0	158.4
Sep	93	78	1.18	198.2	0.0	190.0	SE	19.1	12.7	80	233.4	88.4	0.0	0.0	0	233.4
Oct	89	78	0.93	310.1	0.0	97.5	SE	28.8	17.2	135	366.2	105.4	0.1	1.3	0	366.5
Nov	87	79	0.71	430.9	0.0	24.6	SE	26.7	16.3	137	258.9	69.9	8.8	19.8	2	268.4
Dec	85	80	0.61	502.9	0.0	12.7	SE	26.1	16.8	111	250.0	89.6	36.4	29.5	4	284.1
Ann			0.88	4110.4	0.5	1148.2	SE	19.9	14.8	137	2369.0	105.4	151.7	39.9		2523.2

AVERAGE FROST DATA
Last frost (spring) May 11
First frost (fall)   October 15
Frost-free period   156 days

CLIMATIC EXTREMES
Wettest Month on Record	728.7 mm	October 1974
Highest Temperature on Record	28.7°	August 14, 1977
Lowest Temperature on Record	-24.4°	January 4, 1965

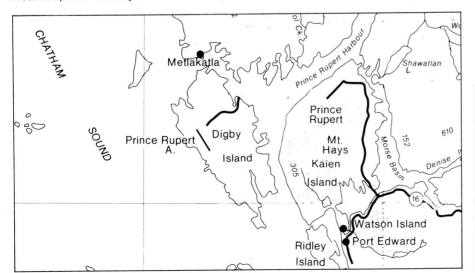

STATION INFORMATION
PRINCE RUPERT   54°18'N   130°26'W   52 m
The airport is located on the western side of Digby Island, 14 km west of the British Columbia mainland. The city of Prince Rupert is situated on the northwest coast of Kaien Island, 8 km east-northeast of the airport. The terrain throughout Digby Island is somewhat hilly, although no significant elevations are attained. This is in sharp contrast to the rugged, mountainous terrain that characterizes the areas to the northeast, east and southeast on the British Columbia mainland and the nearby islands.

Yellowknife, Northwest Territories

LONG-TERM AVERAGES AND EXTREMES OF CLIMATE DATA

	TEMPERATURE (°C)					AVERAGE NUMBER OF DAYS WITH								BRIGHT SUNSHINE	SUNRISE/SUNSET	
	AVERAGE			EXTREMES						Freezing						
	Midafternoon	Early Morning	Daily	Highest	Lowest	Winds (>63 km/h)	Hail	Thunder	Fog	Temp.	Precip.	Rain	Snow	(hours)	Local Standard Time (h:min) on the 15th of Month	
Jan	-24.7	-33.0	-28.8	2.8	-51.1	*	0	0	2	31	1	*	12	44.0	9:49	15:45
Feb	-20.3	-29.8	-25.1	6.1	-51.1	0	0	0	1	28	1	*	10	102.3	8:29	17:16
Mar	-13.1	-24.7	-18.9	8.9	-43.3	0	0	0	1	31	1	*	10	195.3	6:59	18:36
Apr	-1.2	-12.5	-6.9	20.3	-40.6	*	0	0	1	28	1	1	6	266.4	5:16	20:01
May	10.0	-0.1	5.0	26.1	-22.8	*	*	*	1	14	*	4	2	333.6	3:42	21:26
Jun	17.8	8.0	12.9	30.0	-4.4	*	*	1	1	1	*	6	*	394.6	2:41	22:35
Jul	20.7	11.8	16.3	32.2	0.6	*	*	2	*	0	0	9	0	382.1	3:13	22:14
Aug	18.1	10.1	14.1	30.2	0.6	*	*	2	1	0	0	10	0	287.6	4:37	20:48
Sep	9.9	3.5	6.7	26.1	-8.3	*	*	*	2	7	*	9	2	152.0	6:00	19:06
Oct	1.2	-4.4	-1.6	18.3	-28.9	*	0	0	5	25	3	6	9	56.2	7:21	17:27
Nov	-10.1	-18.0	-14.1	7.8	-44.4	*	0	0	4	30	5	1	16	41.7	8:50	15:55
Dec	-19.9	-28.0	-24.0	2.8	-48.3	0	0	0	2	31	1	*	15	20.8	10:01	15:04
Ann	-1.0	-9.8	-5.4	32.2	-51.1	*	*	5	21	226	13	46	82	2276.6		

	HUMIDITY			DEGREE-DAYS (°C)			WIND				PRECIPITATION					
	Relative (%)		Vapour Pressure kPa								RAINFALL (mm)		SNOWFALL (cm)			
	6 a.m. Average	3 p.m. Average		Heating (<18°)	Cooling (>18°)	Growing (>5°)	Prevailing Direction	% Frequency	Average Speed (km/h)	Extreme Gust (km/h)	Average	Wettest Day	Average	Snowiest Day	Average Snow on Ground	Total (mm)
Jan	72	71	0.07	1452.1	0.0	0.0	NW	15.5	13.7	105	T	0.3	15.5	13.0	36	13.3
Feb	73	69	0.09	1218.0	0.0	0.0	E	14.6	13.5	80	T	0.2	13.1	9.1	41	11.2
Mar	73	64	0.13	1144.9	0.0	0.0	E	13.2	14.7	64	T	0.5	14.4	10.9	41	12.4
Apr	77	61	0.30	746.8	0.0	3.2	E	13.1	16.6	89	1.8	8.6	9.8	12.4	13	10.3
May	72	51	0.54	404.6	0.1	68.9	E	12.0	17.0	87	13.6	34.0	3.7	11.2	0	17.2
Jun	67	45	0.84	157.7	4.9	239.1	S	13.4	17.2	79	16.6	20.1	0.2	3.0	0	16.8
Jul	72	48	1.10	70.3	16.7	349.4	S	12.4	16.1	85	33.8	41.1	0.0	T	0	33.8
Aug	82	55	1.09	126.6	6.4	282.8	S	10.1	15.6	80	44.0	82.8	0.0	T	0	44.0
Sep	86	64	0.76	338.8	0.1	80.4	E	11.6	16.6	105	26.6	24.1	3.6	15.2	1	30.5
Oct	87	78	0.49	606.7	0.0	3.3	E	14.9	17.0	84	12.8	35.6	23.1	16.0	7	34.5
Nov	82	80	0.22	962.1	0.0	0.0	E	14.9	15.2	89	0.9	7.1	30.0	14.7	22	24.5
Dec	75	74	0.10	1301.0	0.0	0.0	E	15.1	13.1	80	0.1	1.5	22.0	15.7	30	18.2
Ann			0.48	8529.6	28.2	1027.1	E	12.5	15.5	105	150.2	82.8	135.4	16.0		266.7

AVERAGE FROST DATA
Last frost (spring) May 27
First frost (fall) September 16
Frost-free period 111 days

CLIMATIC EXTREMES
Wettest Month on Record	141.4 mm	August 1969
Highest Temperature on Record	32.2°	July 9, 1964
Lowest Temperature on Record	-51.2°	February 4, 1947

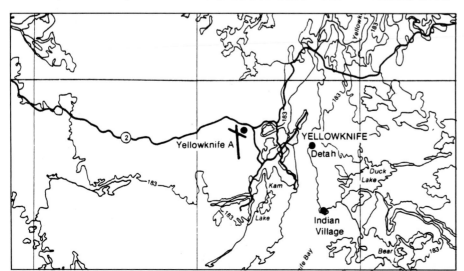

STATION INFORMATION
YELLOWKNIFE 62°28'N 114°27'W 205 m
The airport is situated south of the Makenzie Highway (HWY 2), 4 km west-northwest of the city of Yellowknife and 5 km northwest of Yellowknife Bay on the North Arm of Great Slave Lake. Numerous lakes and marsh areas dot the landscape within a 20-km radius of the site. The land surface consists mainly of bare rock; however, stunted trees and other vegetation do manage to grow in what soil exists, mostly in the depressions around lakes.

Whitehorse, Yukon
LONG-TERM AVERAGES AND EXTREMES OF CLIMATE DATA

	TEMPERATURE (°C) AVERAGE			EXTREMES		AVERAGE NUMBER OF DAYS WITH				Freezing				BRIGHT SUNSHINE	SUNRISE/SUNSET	
	Midafternoon	Early Morning	Daily	Highest	Lowest	Winds (>63 km/h)	Hail	Thunder	Fog	Temp.	Precip.	Rain	Snow	(hours)	Local Standard Time (h:min) on the 15th of Month	
Jan	-16.4	-25.0	-20.7	9.0	-52.2	*	0	0	4	31	*	*	12	46.0	9:55	16:24
Feb	-8.3	-18.1	-13.2	11.7	-51.1	*	0	0	1	28	*	*	10	91.0	8:44	17:45
Mar	-2.3	-14.0	-8.2	11.7	-40.6	*	0	0	*	30	0	*	9	153.1	7:20	18:59
Apr	5.6	-5.1	0.3	20.6	-27.8	0	*	0	*	28	0	1	6	229.6	5:44	20:17
May	12.7	0.6	6.7	30.0	-11.7	*	*	1	*	15	0	5	6	259.2	4:19	21:34
Jun	18.4	5.5	12.0	34.4	-2.8	0	*	2	*	2	0	9	9	272.8	3:28	22:33
Jul	20.3	7.9	14.1	32.8	0.0	*	*	2	*	0	0	11	11	250.2	3:55	22:17
Aug	18.4	6.5	12.5	30.0	-4.4	0	*	1	2	1	0	11	11	230.7	5:09	21:01
Sep	12.4	2.6	7.5	26.7	-11.7	*	*	0	2	8	0	10	11	136.5	6:25	19:27
Oct	4.4	-3.1	0.6	18.9	-28.3	*	0	0	2	22	*	4	10	93.4	7:39	17:54
Nov	-5.3	-12.3	-8.8	11.7	-40.6	*	0	0	2	28	*	1	12	58.3	9:00	16:29
Dec	-12.5	-20.7	-16.6	8.3	-47.8	*	0	0	3	31	1	*	13	23.0	10:03	15:47
Ann	4.0	-6.3	-1.2	34.4	-52.2	*	*	6	16	224	1	52	120	1843.8		

	HUMIDITY Relative (%)		Vapour Pressure kPa	DEGREE-DAYS (°C)			WIND				PRECIPITATION RAINFALL (mm)		SNOWFALL (cm)			
	6 a.m. Average	3 p.m. Average		Heating (<18°)	Cooling (>18°)	Growing (>5°)	Prevailing Direction	% Frequency	Average Speed (km/h)	Extreme Gust (km/h)	Average	Wettest Day	Average	Snowiest Day	Average Snow on Ground	Total (mm)
Jan	77	75	0.16	1199.5	0.0	0.0	SSE	15.3	12.9	100	T	0.5	21.3	14.0	32	17.7
Feb	75	68	0.22	882.5	0.0	0.2	SSE	18.6	14.9	106	T	0.4	15.2	10.4	30	13.3
Mar	75	57	0.26	810.9	0.0	0.1	SSE	18.9	14.3	93	T	0.8	16.4	27.2	22	13.5
Apr	72	46	0.37	532.2	0.0	4.8	SE	19.9	14.3	89	0.8	3.8	10.5	16.3	2	9.5
May	68	38	0.50	351.8	0.0	68.6	SE	22.3	14.4	85	10.1	12.4	2.9	12.2	0	12.9
Jun	69	39	0.74	183.1	1.6	208.6	SE	22.3	12.8	90	29.8	30.2	0.9	12.7	0	30.7
Jul	74	45	0.92	124.3	3.4	282.1	SE	22.5	12.4	80	33.9	21.1	0.0	0.0	0	33.9
Aug	80	48	0.89	173.0	1.2	231.4	SE	21.3	12.4	84	37.0	30.7	0.8	8.6	0	37.9
Sep	82	54	0.70	315.2	0.0	90.9	SSE	23.2	13.8	101	25.9	19.6	4.5	21.6	T	30.3
Oct	79	63	0.48	538.2	0.0	10.1	SSE	27.0	16.4	97	6.7	18.3	16.1	12.2	3	21.5
Nov	79	76	0.29	804.4	0.0	0.4	S	22.4	15.7	106	1.1	9.4	23.8	14.6	15	19.8
Dec	78	77	0.20	1073.3	0.0	0.0	S	19.4	14.5	97	0.2	1.8	24.2	27.0	25	20.2
Ann			0.48	6988.4	6.2	897.2	SSE	19.6	14.1	106	145.5	30.7	136.6	27.2		261.2

AVERAGE FROST DATA
Last frost (spring) June 8
First frost (fall) August 30
Frost-free period 82 days

CLIMATIC EXTREMES
Wettest Month on Record — 103.4 mm — August 1974
Highest Temperature on Record — 34.4° — June 14, 1969
Lowest Temperature on Record — -52.2° — January 31, 1947

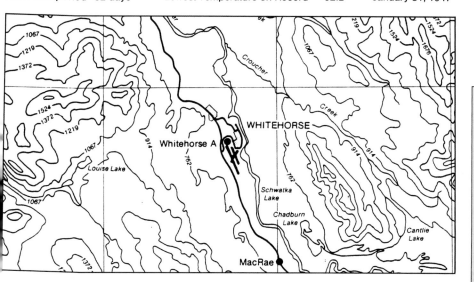

STATION INFORMATION
WHITEHORSE 60°43'N 135°04'W 703 m
The airport is situated adjacent to the steep western escarpment of the Yukon River. The city of Whitehorse lies on the valley floor approximately 60 m below and 400 m east of the airport. The topography of the area consists of mountains ranging in elevation from 1372 to 2134 m and deep narrow valleys, 610 to 914 m. Although the airport is only 130 km north of Lynn Canal, the nearest arm of the Pacific Ocean, a series of mountain ranges acts as a significant rain barrier. To the south of the Airport lies the Coast Mountains with peaks of 2440 m, and lying to the southwest and west a higher barrier is formed by the Saint Elias Mountains with peaks approaching 6100 m.

Resolute, Northwest Territories

LONG-TERM AVERAGES AND EXTREMES OF CLIMATE DATA

	TEMPERATURE (°C)					AVERAGE NUMBER OF DAYS WITH								BRIGHT SUNSHINE	SUNRISE/SUNSET	
	AVERAGE			EXTREMES						Freezing						
	Midafternoon	Early Morning	Daily	Highest	Lowest	Winds (> 63 km/h)	Hail	Thunder	Fog	Temp.	Precip.	Rain	Snow	(hours)	Local Standard Time (h:min) on the 15th of Month	
Jan	-28.4	-35.7	-32.1	-0.8	-52.2	4	0	0	3	31	0	0	5	0.0	PN	PN
Feb	-29.6	-36.8	-33.2	-3.9	-52.0	3	0	0	4	28	*	0	5	17.7	10:01	15:08
Mar	-27.8	-35.0	-31.4	-6.7	-51.7	3	0	0	4	31	0	0	5	145.9	6:52	18:06
Apr	-19.2	-27.0	-23.1	0.0	-41.7	2	0	0	2	30	*	0	7	276.4	3:33	21:07
May	-7.7	-14.1	-10.9	4.4	-29.4	1	0	0	4	31	1	*	9	292.3	MS	MS
Jun	1.7	-2.9	-0.6	13.9	-16.7	1	0	*	9	24	3	3	6	255.8	MS	MS
Jul	6.8	1.4	4.1	18.3	-2.8	1	0	0	11	11	1	8	2	274.4	MS	MS
Aug	4.7	-0.1	2.4	15.0	-8.3	2	0	0	12	17	2	8	5	159.4	0:57	23:52
Sep	-2.9	-7.2	-5.1	9.4	-20.6	1	0	0	6	29	5	1	11	59.1	5:13	19:18
Oct	-12.0	-18.3	-15.1	0.0	-35.0	2	0	0	4	31	1	*	13	23.7	7:59	16:13
Nov	-20.9	-28.0	-24.5	-2.8	-42.8	2	0	0	2	30	0	0	7	0.4	PN	PN
Dec	-25.7	-32.9	-29.3	-6.1	-46.1	3	0	0	1	31	0	0	7	0.0	PN	PN
Ann	-13.4	-19.7	-16.6	18.3	-52.2	25	0	*	62	324	13	20	82	1505.1		

	HUMIDITY			DEGREE-DAYS (°C)			WIND				PRECIPITATION					
	Relative (%)		Vapour Pressure kPa								RAINFALL (mm)		SNOWFALL (cm)			
	6 a.m. Average	3 p.m. Average		Heating (<18°)	Cooling (>18°)	Growing (>5°)	Prevailing Direction	% Frequency	Average Speed (km/h)	Extreme Gust (km/h)	Average	Wettest Day	Average	Snowiest Day	Average Snow on Ground	Total (mm)
Jan	68	68	0.05	1553.1	0.0	0.0	NNW	15.7	22.5	138	0.0	0.0	3.4	3.8	26	3.3
Feb	68	67	0.04	1448.3	0.0	0.0	NNW	16.8	21.3	135	0.0	T	3.1	3.8	27	3.0
Mar	67	67	0.04	1531.0	0.0	0.0	NNW	15.7	20.8	108	0.0	0.0	3.1	3.8	29	3.0
Apr	71	72	0.08	1234.1	0.0	0.0	NNW	13.0	19.7	138	0.0	T	6.5	6.9	31	5.9
May	81	81	0.24	895.5	0.0	0.0	NNW	11.4	21.4	119	T	0.5	9.2	8.6	32	8.1
Jun	88	84	0.51	557.7	0.0	2.4	NNW	11.2	21.7	109	5.3	14.5	7.0	8.4	5	12.1
Jul	87	80	0.68	429.4	0.0	22.3	W	9.8	20.8	108	19.1	20.6	3.3	9.4	T	22.5
Aug	91	85	0.64	485.2	0.0	8.2	SE	11.1	20.8	113	24.6	25.1	6.7	7.9	1	31.1
Sep	89	87	0.39	692.2	0.0	0.1	N	15.6	24.0	100	3.7	15.2	15.3	13.2	9	18.0
Oct	82	82	0.19	1027.7	0.0	0.0	NNW	13.4	23.4	122	T	0.5	14.8	9.4	21	13.8
Nov	73	72	0.08	1273.9	0.0	0.0	NNW	13.8	21.2	158	0.0	0.0	6.1	5.1	22	5.7
Dec	69	69	0.05	1465.7	0.0	0.0	NNW	14.5	20.5	132	0.0	0.0	5.3	5.6	25	4.9
Ann			0.25	12593.8	0.0	33.0	NNW	13.0	21.5	158	52.7	25.1	83.8	13.2		131.4

AVERAGE FROST DATA

Last frost (spring) July 10
First frost (fall) July 20
Frost-free period 9 days

CLIMATIC EXTREMES

Wettest Month on Record	77.4 mm	August 1981
Highest Temperature on Record	18.3°	July 21, 1962
Lowest Temperature on Record	-52.2°	January 7, 1966

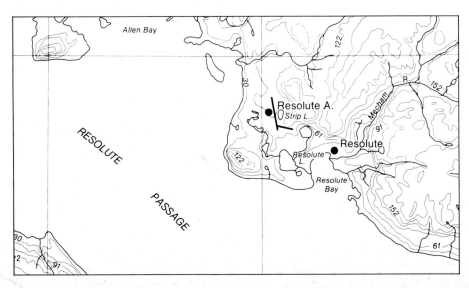

STATION INFORMATION

RESOLUTE 74°43'N 94°59'W 67 m

The airport is situated on a peninsula on the southern tip of Cornwallis Island, 4 km north-northwest of the settlement of Resolute in the Northwest Territories. Resolute Passage, at its closest point, is 3 km west of the station. The terrain, which is gently rolling in character, slopes down to the south and west, toward the sea. There is a sharp ridge of hills about 1 km to the northeast, which rises about 120 m to an upland plateau with rolling slopes.